AF386247

Network-aware Source Coding and Communication

An introduction to the theory and techniques for achieving high-quality network communication with the best possible bandwidth economy, this book focuses on network information flow with fidelity. Covering both lossless and lossy source reconstruction, it is illustrated throughout with real-world applications, including sensor networks and multimedia communications. Practical algorithms are presented, developing novel techniques for tackling design problems in joint network-source coding via collaborative multiple-description coding, progressive coding, diversity routing, and network coding. With systematic introductions to the basic theories of distributed source coding, network coding, and multiple-description coding, this is an ideal self-contained resource for researchers and students in information theory and network theory.

Nima Sarshar is an Associate Professor of Software Systems Engineering at the University of Regina, Canada. The recipient of best paper awards at IEEE P2P 2004 and SPIE VCIP 2008, his research interests include network communication of multimedia signals, large-scale distributed computing, and P2P computing.

Xiaolin Wu is a Professor in the Department of Electrical and Computer Engineering at McMaster University, Canada. His research interests include multimedia signal processing and communications, data compression, and visual computing. He is an IEEE Fellow and currently serves as an Associate Editor of *IEEE Transactions on Image Processing*.

Jia Wang is an Associate Professor in the Department of Electrical Engineering at Shanghai Jiao Tong University, China. His research interests include multi-user information theory and its application in video coding.

Sorina Dumitrescu is an Associate Professor in the Department of Electrical and Computer Engineering at McMaster University, Canada. Her research interests lie in robust image coding, data compression for networks, multiple-description source codes, quantization, and joint source-channel coding.

Network-aware Source Coding and Communication

NIMA SARSHAR
University of Regina

XIAOLIN WU
McMaster University

JIA WANG
Shanghai Jiao Tong University

SORINA DUMITRESCU
McMaster University

Shaftesbury Road, Cambridge CB2 8EA, United Kingdom

One Liberty Plaza, 20th Floor, New York, NY 10006, USA

477 Williamstown Road, Port Melbourne, VIC 3207, Australia

314–321, 3rd Floor, Plot 3, Splendor Forum, Jasola District Centre, New Delhi – 110025, India

103 Penang Road, #05–06/07, Visioncrest Commercial, Singapore 238467

Cambridge University Press is part of Cambridge University Press & Assessment,
a department of the University of Cambridge.

We share the University's mission to contribute to society through the pursuit of
education, learning and research at the highest international levels of excellence.

www.cambridge.org
Information on this title: www.cambridge.org/9780521888400

© Cambridge University Press & Assessment 2011

First published 2011

A catalogue record for this publication is available from the British Library

Library of Congress Cataloging-in-Publication data
Network-aware source coding and communication / Nima Sarshar . . . [et al.].
    p. cm.
  Includes bibliographical references and index.
  ISBN 978-0-521-88840-0 (Hardback)
1. Telecommunication–Data processing. 2. Telecommunication–Traffic. 3. Computer programming.
I. Sarshar, Nima. II. Title.
TK5102.5.N396 2011
005.1–dc23

          2011014524

ISBN   978-0-521-88840-0   Hardback

Contents

1   Introduction   *page* 1
1.1   Network representation of source coding problems   1
1.2   Source coding and communication in networks with more complex
      topologies   3
1.3   Separability of source coding and on-route processing   4
     1.3.1   Lossless communication of a single source   4
     1.3.2   Single source communication to sinks with equal max-flow   4
1.4   More general scenarios   4
     1.4.1   Distributed source coding in arbitrary networks   4
     1.4.2   Lossy communication of a single source in arbitrary networks   5
1.5   Applications and motivations   7
1.6   Network-aware source coding and communication: a formal definition   8
1.7   Organization of this book   9

Part I   The lossless scenario   13

2   Lossless multicast with a single source   15
2.1   Network coding, the multicast scenario   15
2.2   Information multicast with routing only   16

3   Lossless multicast of multiple uncorrelated sources   20
3.1   Multi-unicast problem   20
3.2   Multi-unicast with routing and network coding on directed acyclic graphs   20
     3.2.1   Coding gain in directed networks can be high   21
     3.2.2   Cuts in undirected graphs   22
     3.2.3   Coding gain in undirected networks: Li and Li's conjecture   23
3.3   Concluding remarks   23

4   Lossless multicast of multiple correlated sources   25
4.1   Slepian-Wolf problem in simple networks   25
     4.1.1   Main theorem and its proof   26
     4.1.2   Slepian-Wolf coding for many sources   29
     4.1.3   Slepian-Wolf code design   29

	4.2	Slepian-Wolf problem in general networks	34
	4.3	Concluding remarks	36

Part II The lossy scenario   37

5 Lossy source communication: an approach based on multiple-description codes   39
	5.1	Separating source coding from network multicast	39
	5.2	Beyond common information multicast: rainbow network flow	41
	5.3	Multiple-description coding: a tool for NASCC	42
	5.3.1	Example 5.1	43
	5.3.2	Example 5.2	45
	5.3.3	Design issues	47
	5.4	Rainbow network flow problem	48
	5.5	Code design	50
	5.5.1	MDC using PET	50
	5.5.2	Optimizing code for a fixed rainbow flow	52
	5.5.3	Discrete optimization approaches	53
	5.6	Numerical simulations	54
	5.6.1	Network simulation setup	54
	5.6.2	Effect of the number of descriptions	55
	5.6.3	The effect of network size	55
	5.6.4	The effect of the performance of path optimization algorithms	57
	5.7	Concluding remarks	58

6 Solving the rainbow network flow problem   59
	6.1	Complexity results of the CRNF problem	59
	6.2	A binary integer program for CRNF on directed acyclic graphs	61
	6.2.1	Formulation	61
	6.2.2	The DAG requirement	62
	6.3	Solving CRNF on tree-decomposable graphs	62
	6.3.1	Calculation of the optimal flows	63
	6.3.2	Flow coloring	65
	6.4	Optimal CRNF for single sink	67
	6.5	Concluding remarks	70

7 Continuous rainbow network flow: rainbow network flow with unbounded delay   71
	7.1	Continuous rainbow network flow	71
	7.2	Achievability results	72
	7.3	Concluding remarks	81

8 Practical methods for MDC design   82
	8.1	Overview of MDC techniques	82
	8.2	Optimal design of multiple-description scalar quantizers (MDSQ)	84

	8.2.1	MDSQ–definition and notations	84
	8.2.2	Generalized Lloyd algorithm for optimal MDSQ design	87
	8.2.3	Index assignment	88
8.3		Lattice MDVQ	91
	8.3.1	Preliminaries	92
	8.3.2	Distortion of MDLVQ	94
	8.3.3	Optimal MDLVQ design	96
	8.3.4	Index assignment algorithm	96
	8.3.5	K-fraction sublattice	97
	8.3.6	Greedy index assignment algorithm	98
	8.3.7	Examples of greedy index assignment algorithm	99
	8.3.8	Asymptotically optimal design of MDLVQ	101
	8.3.9	Asymptotical optimality of the proposed index assignment	101
	8.3.10	Optimal design parameters v, N, and K	102
	8.3.11	Non-asymptotical optimality for $K = 2$	107
	8.3.12	A non-asymptotical proof	107
	8.3.13	Exact distortion formula for $K = 2$	108
8.4		S-similarity	111
8.5		Local adjustment algorithm	113
8.6		PET-based MDC	116
	8.6.1	Exact solution for the general case	119
	8.6.2	Fast matrix-search algorithm for convex case	121
8.7		General MDC based on progressive coding	125
	8.7.1	Framework description	125
	8.7.2	G-MDC rate optimization	127
	8.7.3	G-MDC description construction	129
9		**Using progressive codes for lossy source communication**	132
9.1		Lossy source communication with network coding: an introduction	132
9.2		Formulation	134
9.3		Layered multicast with intra-layer network coding	135
	9.3.1	Conditions for absolute optimality of LM with intra-layer network coding	135
	9.3.2	Optimization of layered multicast strategy	138
9.4		Layered multicast with inter-layer network coding	141
	9.4.1	Flow optimization for inter-layer network coding	141
	9.4.2	Network code construction	148
	9.4.3	Performance evaluation	152
	9.4.4	Conclusions	155
10		**Lossy communication of multiple correlated sources**	158
10.1		Simple Wyner-Ziv	158
10.2		The Wyner-Ziv theorem and its proof	159

	10.2.1	Strong typicality and Markov lemma	159
	10.2.2	Proof of the Wyner-Ziv theorem	160
10.3		Wyner-Ziv function for Gaussian sources and binary sources	162
	10.3.1	Gaussian sources	162
	10.3.2	Binary sources	162
10.4		Wyner-Ziv code design	163
10.5		Problems closely related to Wyner-Ziv coding	164
	10.5.1	The (direct) multi-terminal source coding problem	164
	10.5.2	The CEO problem (indirect multi-terminal source coding)	167
10.6		A summary of the network source coding problem of no more than two encoders and decoders	168

References	171
Index	178

1   Introduction

To state intuitively, the question investigated in this book is the following:

How does one communicate one or more source signals over a network from nodes (servers) that observe/supply the sources to a set of sink nodes (clients) to realize the best possible reconstruction of the signals at the clients?

The above question sets the unifying theme for the problems studied in this book, and, as will be made clear in this introduction, contains some of the most important and fundamental problems in information and network communication theory.

1.1   Network representation of source coding problems

Let's start with the observation that even the simplest source coding problems have (perhaps trivial) network representations. Figure 1.1, for instance, shows network representations of the three arguably most fundamental source coding problems. Here, the goal is to communicate a single source signal X, from a single server node s to one or more sinks. Each link of the network has a "capacity" assigned to it, which, when properly normalized, indicates the number of bits that can be communicated over that link, without errors, for every source symbol emitted from X.

Figure 1.1(a) is the simplest source coding problem. The receiver node t receives an R_1 bits encoding of X from the source node s. If X admits a rate-distortion function $D_X(R)$, the reconstruction error at t is at best $d_t = D_X(R_1)$. Of course if X is defined on a finite alphabet, then for large enough R_1, there is a possibility to communicate X losslessly, or without distortion (i.e., $d_t = 0$).

Figure 1.1(b), on the other hand, is the network representation of the progressive source coding problem, when $R_1 > R_2$, while Fig. 1.1(c) represents the network of two-description multiple-description source coding problem. In the latter case, sink node t_1 receives a description of X with rate R_1, while t_2 receives another description of rate R_2. Node t_3 receives both descriptions, and in the terminology of multiple-description coding (MDC), it acts as the *joint decoder*.

These examples correspond to the case of a single source signal X. Same is true for problems involving multiple source signals observed by multiple, usually non-communicating, encoders; problems often called distributed source coding. The problem becomes particularly interesting when the multiple signals in question are correlated. Figure 1.1(a) is the simplest setting of a distributed source coding problem in which two correlated sources X and Y are encoded at two separate (non-communicating) nodes s_1 and s_2, and are communicated to a single receiver node (or decoder), t.

This problem is called Slepian-Wolf distributed source coding [1] when the two sources are on finite alphabets and the goal is lossless communication of both of them to the sink t. Slepian and Wolf found necessary and sufficient conditions on the rates R_1 and R_2 for which lossless communication of both sources X and Y to t is feasible. In the special case where $R_2 = \infty$, i.e., the source Y is losslessly present at the decoder only, the problem of distributed source coding is usually called the Wyner-Ziv problem [69].

There is a large body of literature on theoretical and practical aspects of source coding problems with simple network representations as in Figs. 1.1 and 1.2. These include the complete characterization of achievable rate-distortion regions for many classes of important signals [2], [1] as well as powerful practical coding approaches that

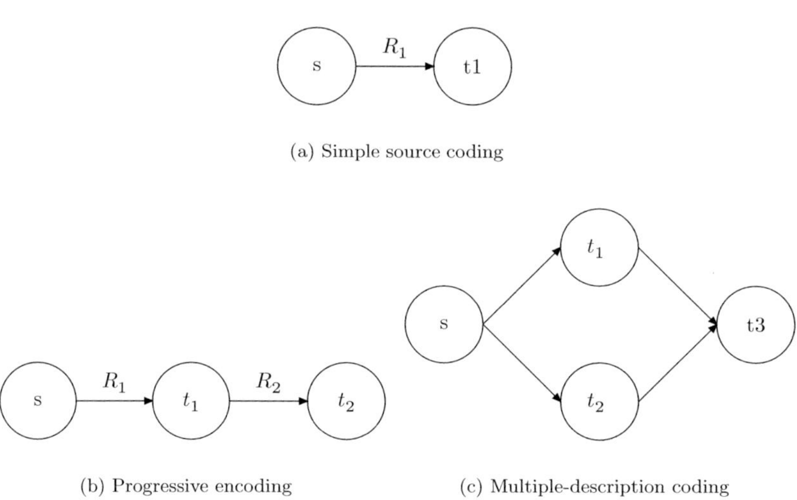

(a) Simple source coding

(b) Progressive encoding   (c) Multiple-description coding

Figure 1.1 Network representation of fundamental source coding problems

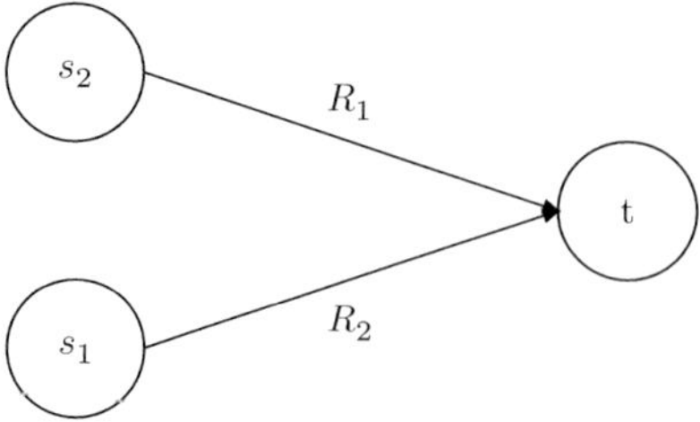

Figure 1.2 The network representation of Slepian-Wolf, Wyner, and Wyner-Ziv problems

perform close to these theoretical limits, some of which will be reviewed throughout this book.

1.2 Source coding and communication in networks with more complex topologies

For networks discussed so far, network structures are very simple. Since the communication links are assumed to be error-free, the network communication aspect of the source coding-and-networking problem is trivial. In particular, once the source or sources are encoded, the encodings are simply passed over the corresponding links. Recent advances in network information flow (e.g., network coding), however, suggest that there is much more to network communication than simple information relay.

This book intends to not only cover some of the above mentioned special cases of source coding in networks, but to go beyond, by exploring problems with networks of complex topologies. Figure 1.3 shows the two simplest network information flow scenarios. In both scenarios, there is a single source to communicate from s to one or more sink nodes, over an arbitrarily complex network. For these arbitrary networks, as schematically depicted in Fig. 1.3, a new dimension enters the problem – that of on-route information processing. As will become clear shortly, in most scenarios, source coding is accompanied by network coding and routing (an aspect that we call on-route processing) in a nontrivial fashion and, thus, in general, the two have to be considered jointly. In other words, optimal utilization of network resources requires a joint consideration of both source coding and on-route processing.

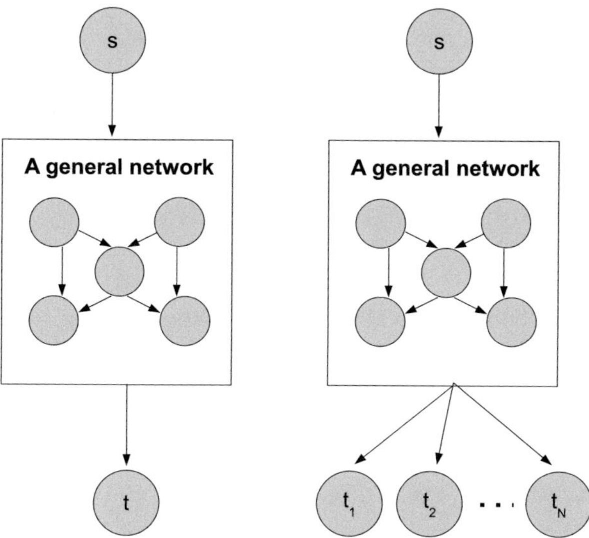

Figure 1.3 Unicast vs. multicast source communication

1.3   Separability of source coding and on-route processing

Before getting to more general scenarios, we first review some of the few cases in which source coding and on-route processing can in fact be separately performed without loss of optimality.

1.3.1   Lossless communication of a single source

If X is defined on a finite alphabet, and the goal is a lossless communication of X from s to all the sink nodes, then source coding and communication can be broken down into two steps without loss of optimality, (1) a source coding step and, (2) a network communication step. The network communication step, however, is fundamentally different in the unicast and multicast scenarios.

Suppose the entropy of the source signal X is $H(X)$. Then, in the unicast scenario, X can be communicated to the single source node t losslessly if and only if the max-flow from s to t is at least $H(X)$. Furthermore, it suffices to perform only routing (in fact simple relaying) in the network communication step. For the multicast scenario, on the other hand, the source X can be communicated from s to sinks, if and only if the max-flow from s to every sink node is at least $H(X)$. Furthermore, network communication, in general, requires re-encoding of information at relay nodes (i.e., network coding).

1.3.2   Single source communication to sinks with equal max-flow

Another special case in which source coding and network communication can be separated without loss of optimality is when a single source X has to be communicated from s to sink nodes with equal max-flow h. In that case, one can separately encode X to rate h and then use network coding (in the multicast scenario) or simple routing (in the unicast scenario) to communicate the source encoding to the sink nodes. If the source admits a rate-distortion function $D_X(R)$, the distortion $D_X(h)$ is achievable at all sink nodes. It is easy to verify that this is the smallest achievable distortion, given that the max-flow to each sink node is h. These results are reviewed in more detail in Chapter 2.

1.4   More general scenarios

It turns out that source coding and network communication are non-separable in most other scenarios, some important cases of which are reviewed next.

1.4.1   Distributed source coding in arbitrary networks

The Slepian-Wolf network settings in Fig. 1.2 can be generalized to the case of an arbitrary network, as in Fig. 1.4. If X and Y are on finite alphabets, necessary and sufficient conditions under which they can be losslessly communicated to the sink nodes $t_1, t_2, \ldots, t_n$ have been found recently [3].

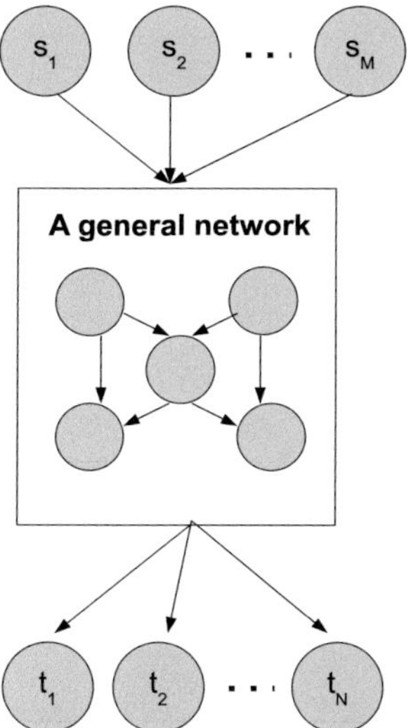

Figure 1.4 Slepian-Wolf coding in general networks

One important question discussed in [4] is whether distributed source coding can be separated from network communication for Slepian-Wolf problem in arbitrary networks. What happens if one encodes X to R_1 bits, and multicasts it from s_1 to all receivers and then encodes Y into R_2 bits and multicasts it to all sinks from s_2? It turns out that this separation strategy is suboptimal in general. More precisely, for many cases, it is impossible to losslessly communicate X and Y to all the sinks unless distributed source coding and network coding are done jointly. This warrants the study of a new class of codes, which can be called joint network source codes (JNSC).

1.4.2 Lossy communication of a single source in arbitrary networks

As we saw earlier, when there is only a single sink, or when the max-flow to all sink nodes is the same, source coding and on-route processing can be separated without loss of optimality.

However, the nature of the problem changes drastically when the set of sink nodes has heterogeneous flow properties (i.e., they don't have equal max-flows). In this case, a joint consideration of source coding and on-route processing is necessary even when only a single source is transmitted in the network. The scenarios in Fig. 1.5 illustrate how source coding and on-route processing (here in particular network coding) can become entangled in a complex way, as explained below.

Figure 1.5(a) illustrates the case where the source X has to be communicated only to the nodes $5, 6$. The max-flow into both nodes 5 and 6 is 2. Thus, as stated before, one can optimally encode X into a source code stream of rate 2, break the stream into two sub streams a and b, each of rate 1, and communicate them to nodes 5 and 6. Note that network coding (i.e., bitwise XOR operation on streams a and b) is necessary at node 3. Methods and results in network coding are briefly reviewed in Part I of this book.

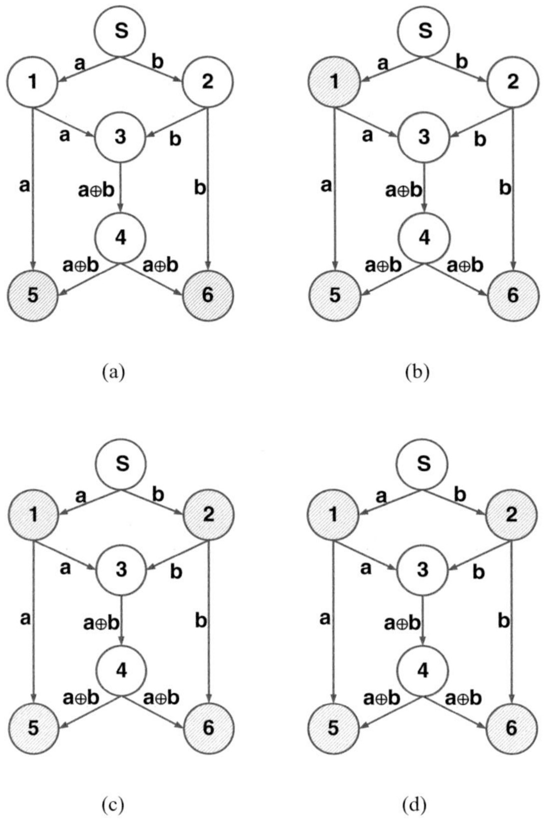

(a)   (b)

(c)   (d)

Figure 1.5 The Butterfly network for which all links have capacity one: When the sink nodes are, (a) 5 and 6, (b) 1, 5, and 6, (c) 1, 2, 5, and 6, (d) 4, 5, and 6

But, what if nodes 1, 5, and 6 constitute the set of sink nodes, as in Fig. 1.5(b)? In this case, the max-flow is 1 for node 1 and is 2 for nodes 5 and 6. The optimal strategy is now to progressively encode X into a stream of rate 2. Take the first portion of this stream to make another stream a of rate 1 and make the rest of the stream into another stream named b, again of rate 1. Network coding should be used to get both a and b to nodes 5 and 6 and stream a to node 1. Note that node 2 gets the stream b only. Since this is the second portion of a progressively encoded source code stream, node 2 will not be able to reconstruct X at all (but this is OK, since node 2 is not a sink node). This "layered" coding and communication strategy is reviewed in Chapter 9.

Figure 1.5(c) is yet another scenario, in which nodes 1, 2, 5, and 6 are sink nodes. In this case, the most general strategy is to encode X into two multiple-description source code streams each of rate 1 (name them streams a and b). Then use network coding to communicate both a and b to nodes 5, 6. Node 1 will receive only the description a and node 2 will only receive b. The use of multiple-description codes for efficient source multicast is reviewed in Chapters 5 and 6, while practical methods for designing multiple-description code streams are reviewed in Chapter 8.

As Fig. 1.5(a) to Fig. 1.5(c) suggest, sometimes, it is possible to break (without the loss of optimality) the task of source coding and on-route processing into a proper concatenation of well understood source coding operations (e.g., progressive coding or MDC), followed by network communication techniques (e.g., network coding or routing). This breakdown, however, cannot be done blindly. In other words, the choice of source encoding and network communication strategies should be made jointly.

In most other cases, however, full, joint consideration of source coding and on-route processing is required. When nodes 4, 5, and 6 are sinks, for instance, none of the above strategies is necessarily optimal (Fig. 1.5(d)).

The above examples suggest that a separate, formal treatment of the problem of source communication in networks is required. In this book, we will gather all these problems under the same umbrella, that of *Network-aware Source Coding and Communication* or NASCC. We will review a wide spectrum of new and old results related to different instances of the NASCC problem.

1.5   Applications and motivations

Our model of networks adopted in this text, in many ways, is an abstraction of computer networks, and is consistent with layered design of today's network protocols. The network is modeled as a graph of interconnected nodes that can communicate with rates constrained to the capacity of network links, i.e., the topological structure of the network is explicitly taken into account.

Our model is particularly relevant to the Internet at the router level and to overlay Peer-to-Peer (P2P) networks. As such, the immediate and by far the largest application domain of this research is real-time multimedia streaming over the Internet. But all the results of this book are valid in any other application area in which a data source (e.g., a physical measurement) has to be relayed to one or more receivers over an underlying network, an important instance of which is signal communication in sensor networks. For the clarity and concreteness of the presentations, this book will limit its discussions on applications to networked multimedia communications.

Real-time multimedia communication spans a wide range of applications, including digital TV and radio broadcasts over the Internet (e.g., IP-TV [5]), video conferencing, video on demand (VoD), distant education, telemedicine, voice over IP (VoIP), online computer games, virtual whiteboards, security and surveillance modules, and many others. Current estimates show that real-time multimedia traffic generated by real-time streaming and VoIP only, accounts for more than 21 percent of the overall Internet traffic in Europe [6], a share that is expected to increase dramatically with the advent of IP-TV technologies. Multicast applications are arguably the most resource-intensive multimedia applications on the web. Most radio stations as well as hundreds of TV channels now stream their live programs on the web. In February 2006, 148 million users listened to radio stations streamed through Shoutcast.com [7] alone.

The NASCC problem studied in this book investigates optimal utilization of bandwidth resources for multimedia multicast applications. On the theoretical side, the study of the NASCC problem, we believe, can fundamentally change our view of signal communication in networks, in much the same way as network coding (discussed in the next chapter) has changed our view of network information flow in the past few years.

1.6   Network-aware source coding and communication: a formal definition

The network model considered in this book is similar to now standard models in network information flow theory [8]. In particular, the model of a network is very close to one's intuition of a computer network: an interconnected set of nodes, each capable of processing and making decisions, which can reliably communicate over their connections provided that the capacities of all connection links are respected.

We are interested in designing a networked communication system to communicate a source signal from a set of source nodes (servers) to a set of sink nodes (clients), so that the source signal can be reconstructed with the best average quality at all the sink nodes. Unlike most frameworks of network information flow, the reconstruction of the source does not need to be perfect. In fact, for most real valued multimedia signals, perfect reconstruction is not necessary or even possible (the digitization process is already lossy in nature). Another major difference is that the quality of the source reconstruction and the input data used for such reconstruction do not need to be the same for all the sinks.

In the version of the problem discussed in this book, it is assumed that the source has been compressed off-line and has been deployed at the server nodes in advance. Again, this formulation is from a networked multimedia application point of view, where a multimedia content (e.g., a video clip) is encoded off-line and is deposited at one or more server nodes in the network before the communication starts. Such an assumption is of course not necessary when there is only a single server node in the network, a case that in fact includes some of the most interesting scenarios, such as live media streaming.

At the time of presentation, the content is *streamed* to one or more users. The communication capacity of the links limits the amount of information that can be communicated from node to node and hence the quality of the reconstruction of the source signal at the sink nodes. We call the problem of finding the strategy that maximizes the overall quality of the signal reconstruction at sink nodes, Network-aware Source Coding and Communication, and it is formally defined next.

Problem formulation:
Formally, the Network-aware Source Coding and Communication (NASCC) problem is defined by the following elements:

- A directed graph $G\langle V, E \rangle$ with nodes set V and edge set $E \subset V \times V$.
- A number of, possibly correlated, sources $X_1, X_2, \ldots, X_K$ over some common alphabet Γ. Of particular interest is the case where $\Gamma = \mathbb{R}^N$, for some N.

- A function $R = E \rightarrow \mathbb{R}^+$ that assigns a capacity $R(e)$ to each link $e \in E$. Bandwidths are normalized with the source bandwidth; therefore, $R(e)$ is expressed in units of bits per source symbol.
- $S_i, T_i \subseteq V$, for $i = 1, 2, \ldots, K$ that denote the set of server and sink nodes respectively for source X_i. We let $S = \cup_i S_i$, $T = \cup_i T_i$ denote the set of all server and sink nodes. The server nodes observe, encode, and communicate X_k in the network. Server nodes are not able to directly communicate (or collaborate) in the encoding process.

Throughout, we assume each source X admits a rate-distortion function $D_X(\cdot)$ under some family of distortion measures. For the most part, we also assume the source X is progressively refinable.

Nodes can communicate with neighbor nodes at a rate specified by the capacity of the corresponding link. $R(e)$, therefore, specifies that an average of $R(e)$ bits can be successfully communicated over link e per source symbol emitted from X.

The task is to communicate the source X_i from the server nodes in S_i and reconstruct X_i at the sink nodes in T_i. Throughout this book, $|\cdot|$ denotes the cardinality of a finite set. Distortion vectors $\mathbf{d}_k = (d_t, t \in T_k) \in \mathbb{R}^{|T_k|}$ for $K = 1, 2, \ldots, K$ are said to be simultaneously achievable if X_k can be reconstructed with an average distortion of d_t at sink node $t \in T_k$ by using a coding scheme that respects the capacity constraints on the links, i.e., the rate of information per source symbol communicated over e is no greater than $R(e)$.

Unlike classical point to point, or even multi-terminal information theory, it proves extremely hard to completely characterize the most general class of possible codes. Therefore, just as in [8], we need to leave the details of the code unspecified.

A number of considerations about the above formulation are due. For clarity, let's assume that there is only one source X, with one set of receivers T.

- A theoretically intriguing problem is how to characterize the set of all achievable distortion vectors $\mathbf{d} \in \mathbb{R}^{|T|}$. Note that this problem includes the usual lossless network coding problem as its special case, if the source alphabet Γ is finite.
- An equivalent problem, which is more relevant practically, is that of finding a coding scheme to minimize a weighted average distortion $\bar{d}(\mathbf{p}) = \sum_t p_t d_t$ for a weighting vector of Lagrangian multipliers $\mathbf{p} = (p_t; t \in T)$. This formulation is mostly considered in this book.

The remainder of this book is a systematic review of the known results and recent developments in dealing with the NASCC problem according to the taxonomy presented in the introduction.

1.7   Organization of this book

Finding the most general source coding-network communication strategy remains an open problem, with little hope for a solution. Even some of its simplest special cases

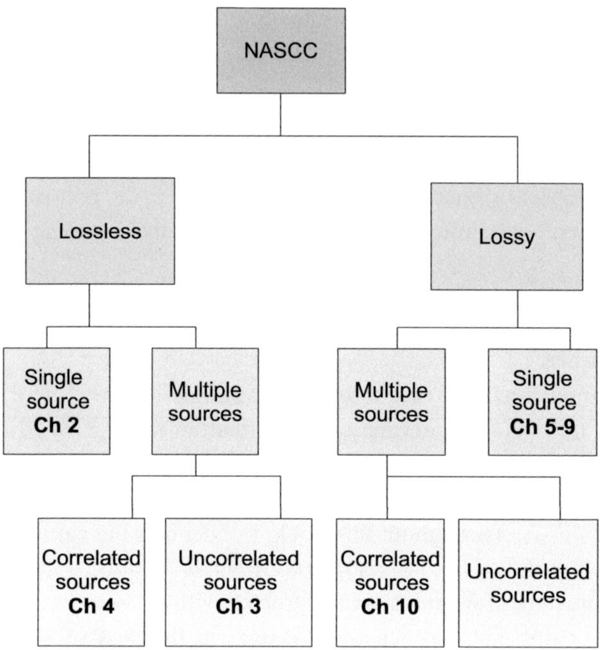

Figure 1.6 A taxonomy of source coding and communication problems in networks and their relation to the chapters of this book

(e.g., *n*-description source encoding) are known to be notoriously hard problems. In this book, however, we adopt a pragmatic approach. We investigate the solutions to this problem that use well-understood source coding techniques (e.g., progressive encoding, MDC, Slepian-Wolf, or Wyner-Ziv coding) along with optimized network communication strategies (e.g., optimized routing, relaying, or network coding), to come up with strong and *practical* solutions.

Figure 1.6 is a taxonomy of NASCC problems and their relationship to the chapters of this book. A brief description of the chapters is given below.

- Chapter 2: Part I of this book, which starts with Chapter 2, deals with lossless communication of sources in networks. Chapter 2 is concerned with the case with only one source node. We will consider scenarios where network coding is and is not allowed. There are several excellent tutorials and textbooks already available on network coding. As such, this chapter is intended to review recent results with an emphasis on algorithmic and complexity perspectives of designing optimal information delivery mechanisms.

- Chapters 3 and 4: In these chapters, we extend the discussions from Chapter 2 to the case of multiple sources. We consider the cases where information sources are correlated and uncorrelated, as well as unicast and multicast scenarios. Chapter 3 covers the case of independent source signals. Results and algorithms when network coding is and is not allowed are reviewed. Discussions will include Li and Li's conjectures (when network coding is allowed) and multi-commodity flow problems (when network coding is not allowed). Chapter 4 reviews known results for multiple correlated sources. Our discussion covers the Slepian-Wolf problem and its generalization to arbitrary networks and multicast scenarios.

- Chapter 5: This chapter is the start of the second part of the book and deals with the problem of lossy source communication in networks. Since lossy reconstruction is now allowed, the rate-distortion optimization framework enters the picture. We will discuss and motivate the problem through examples, which help the reader appreciate the fundamental difference between the lossy and lossless scenarios.
- Chapter 6: In this chapter, we will review the Rainbow Network Flow (RNF) problem in detail. RNF is the problem of RD-optimal distribution of multiple-description codes in arbitrary networks and serves as the building block of some of the practical solutions to lossy NASCC.
- Chapter 7: In this chapter, we investigate the problem of optimization of multiple-description code designs and their network delivery strategies. We will review a powerful practical solution to lossy source delivery in networks using the tools developed in this chapter and Chapter 6.
- Chapter 8: Other powerful lossy source delivery strategies are reviewed in Chapter 8. These include, in particular, layered multicast of progressive codes with network coding.
- Chapters 9 and 10: Finally, the generalization of the methods in Chapters 6–8 to the case of multiple source signals are investigated. In particular, we review Wyner-Ziv problems and their generalizations to the case of arbitrary networks, as well as routing with side information.

Part I

The lossless scenario

2 Lossless multicast with a single source

We start by perhaps the simplest variation of the NASCC problem, namely that of lossless information multicast. This problem has been the source of much interest and studies in the past several years, as ignited by the original work on network coding [8]. We are not going to review the large and exciting body of work on lossless multicast in this book in great detail. However, we will review some of the basic concepts in order to put this book in a proper historical and comparative context. We refer the reader to an array of excellent new books on network coding for further reading [57, 58].

2.1 Network coding, the multicast scenario

In the notations of this book, the network information flow problem introduced by Ahlswede *et al.* in [8] can be defined, for the case of a single information source, with the following elements.

- A directed graph $G\langle V, E \rangle$ with node set V and edge set $E \subset V \times V$.
- A function $R : E \to \mathbb{R}^+$ that assigns a capacity $R(e)$ to each link $e \in E$.
- An information source I that generates information at a server node $s \in S$ at a rate h bits per time unit.
- A set of sink nodes $T \subseteq V$ that are interested in receiving this information.

The multicast demand h is said to be admissible with capacity constraint function R, or equivalently (G, S, T, R, h) is said to be admissible if there exists a coding scheme that satisfies the multicast requirement rate h and respects the capacity constraints on all links $e \in E$. In other words, the number of bits communicated over link e should not exceed $R(e)$ bits per second. Evidently, the information multicast problem, as formulated above, is a special case of the NASCC problem with only one source.

At the heart of network coding is the following celebrated theorem [8] on the maximum admissible multicast demand h.

THEOREM 2.1 (Ahlswede *et al.*) *(G, S, T, R, h) is admissible if and only if for any $t \in T$, there exists a max-flow of at least h into node t.*

If *a single* flow exists that can deliver a flow of at least h to all nodes $t \in T$ simultaneously, then the above theorem is straightforward. In particular, a simple routing algorithm can deliver the demand to all nodes. The problem is that a max-flow to a node

t is not necessarily a max-flow to another node $t' \neq t$. The key technique to achieve the promise of Theorem 2.1 is called network coding. Network coding consists of algebraic mixing of the information bits at intermediate nodes in the network before communicating them to other nodes.

Recent years have seen a tremendous amount of research on network coding techniques. Notably, linear coding over a large enough finite field was shown to be sufficient to achieve the result in Theorem 2.1 [9]. As such, linear network coding has been extensively studied and efficient linear network coding algorithms have been proposed [10–12].

Network coding can in general increase the multicast "capacity" of a network. In Part II of this book, we will examine the suitability of network coding as a tool for dealing with the general NASCC problem.

Network coding is readily used in overlay and P2P networks where the peers are capable computers and hence able to perform complex coding. In other applications, such as router level communication in the Internet, routers are only able to duplicate and forward data, i.e., to multicast. Obviously, the data duplication and multicast can improve bandwidth utilization by not sending duplicates of the same data through a link. The question is whether and how we can realize the full potential of multicast by routing data flows from servers to clients in a way that maximizes a collective information fidelity metric.

When there is only a single server node serving a single client, the ability to duplicate information at relay nodes does not increase the maximum amount of information that can be communicated to the client. In this scenario, maximizing the rate of information flow reduces to the usual max-flow problem from the server to the client.

Finding an optimal routing strategy becomes more interesting, and much more complex, in multicast applications with more than one client. The classic multicast flow problem is about maximizing the amount of *common information* that can be communicated from a server node to a group of clients (multicast group).

We start by reviewing the common information multicast flow problem. Later in Section 5.2, we will motivate a new formulation of the problem which removes the constraint that the data streams received by all clients have to be the same. In this new formulation, called the Rainbow Network Flow problem, different clients can receive different data streams from the server.

2.2   Information multicast with routing only

Network coding requires transcoding of information at relay nodes. When such transcoding is not possible, as is currently the case with Internet routers, one may seek to maximize the rate of common information with routing only. In practice, one needs to answer this question in at least two different scenarios:

The case of integral routing: delay constraints

In most real-time multimedia applications, data are encoded and packetized into a number of data "streams." In the current practice, these streams mainly correspond to different encodings of the multimedia source, which are intended for clients with different bandwidths.

Each data stream is usually a concatenation of blocks of data that have to be decoded together. In video coding, for instance, a group of frames is usually encoded together into a block. For uninterrupted playback, all data packets corresponding to each block have to be present at the client (decoder) before the presentation deadline of that block. Therefore, it is highly desirable in practice to ensure that the data packets corresponding to one stream are routed along the same network path. In other words, the integrity of the data streams has to be respected in the routing process. This form of the flow, which does not split data streams at relay nodes, is sometimes called *integral flow*.

This scenario is best illustrated in Fig. 2.1. Suppose there are two real-time multimedia streams generated at node 1 at a rate of C bits per second each. To be specific, lets assume that node one compresses every second of a multimedia signal into two streams of C bits each. The goal is to communicate these two streams to the sink nodes $T = \{4, 5\}$. The capacity of each link is assumed to be C bits per second. With delay constraint set to at most 1 second, it is easy to see that at most one of the two streams can be communicated to both nodes within 1 second *provided that the streams are required to flow over the same path as a whole*. One way to do this is depicted in Fig. 2.1(a), where we have identified the flow of the two streams with dashed and solid lines.

With such stringent delay and integrity constraints, some links of the network are left unused. To make sure that both streams can reach both nodes, a delay of at least two seconds is required. One can simply send the first stream in the first second as in Fig. 2.1(a) and repeat the same routing in the next second for the other stream. This, of

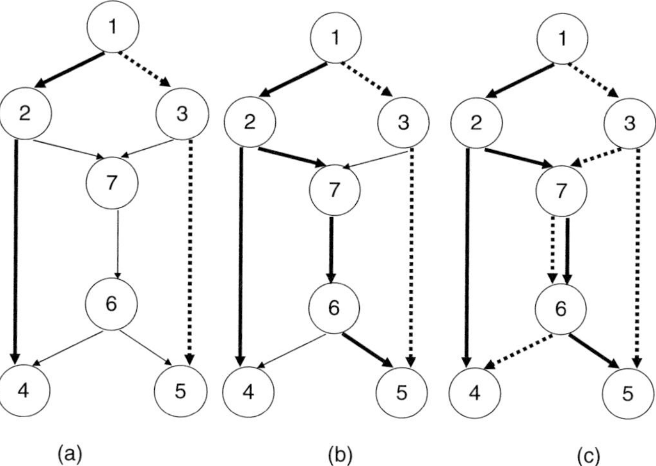

Figure 2.1 Common information multicast from server node 1 to nodes 4, 5 with delay constraint one. (a) Each stream can be communicated over two separate paths. (b) Both streams can be received by node 5 in the first round, while only one stream is received by node 4. (c) In the next round, the other stream is streamed to node 4. In either strategy, only one common information stream per unit time can be streamed to both nodes

course, is not the only possible routing strategy. Fig. 2.1(b) shows an alternative routing in which node 5 receives both streams in the first second while node 4 receives only the solid stream. In the next second, the dashed stream can be communicated to node 4 through $1 \rightarrow 3 \rightarrow 6 \rightarrow 4$ (Fig. 2.1(c)).

The case of fractional routing: relaxing the delay constraint

In the above example, delay constraints, as well as the requirement that the streams should not be split in relay nodes, might limit the volume of data that can be communicated into the sink nodes. In fact, the throughput is only C bits per second, which can be improved to $1.5C$ when the delay constraint is relaxed. This form of routing, which essentially treats data as a continuous medium, is usually called fractional routing.

Without delay constraint (as is the fundamental realm of information theory), one can often communicate a larger rate of common information to sink nodes using, e.g., time-sharing. Take the stream generated at node 1 during a period of n seconds. One can use time-sharing with the routing in Fig. 2.1(b) and its reflection (where solid and dashed descriptions have been interchanged) to communicate at least $\left\lfloor \frac{3}{4}n \right\rfloor C$ bits of common information from each description to the two nodes 4, 5. This corresponds to a total flow of $1.5C$ bits per second to each node for large n.

Maximizing the flow of common information in either of the two scenarios considered above can be formulated as maximum packing of Steiner trees, a problem defined below.

Given a network $G\langle V, E \rangle$ and two subsets $S, T \subset V$, a Steiner tree $ST(s)$ is (for the sake of our presentation) defined as a directed rooted tree with a root $s \in S$ that contains all terminal (or sink) nodes $t \in T$. Let $\text{ß}(G, S, T)$ be the set of all such Steiner trees. For each Steiner tree $\mathcal{I} \in \text{ß}(G, S, T)$ and each link $e \in E$, let $\partial(\mathcal{I}, e)$ be equal to one if e is an edge of $\mathcal{I}$ and zero otherwise.

The maximum common information that can be communicated from nodes in S to nodes in T can be found by solving the so-called maximum Steiner Tree Packing (STP) problem. A $STP(J, \mathcal{J})$ consists of a subset J of $\text{ß}(G, S, T)$ and an assignment $\mathcal{J}$ of weights $\mathcal{J}(\mathcal{I})$ to the trees $\mathcal{I} \in J$.

A $STP(J, \mathcal{J})$ is admissible if it respects the capacity constraint on all edges, i.e.,

$$\forall e \in E, \quad \sum_{\mathcal{I} \in J} \partial(\mathcal{I}, e) \mathcal{J}(\mathcal{I}) \leq R(e). \tag{2.1}$$

The maximum STP problem is that of finding an admissible $STP(J, \mathcal{J})$ that maximizes the common information flow,

$$\sum_{\mathcal{I} \in J} \mathcal{J}(\mathcal{I}).$$

For the case with unbounded delay, the weights $\mathcal{J}(\mathcal{I}) \geq 0$ can be any positive real numbers (known as fractional packing in the literature). To take delay into account, and ensure the integrity of data streams, one has to force $\mathcal{J}(\mathcal{I})$ to be a positive integer (called integer packing). When delay n can be tolerated, each link capacity should be multiplied by n before solving the maximum STP problem on integer weights.

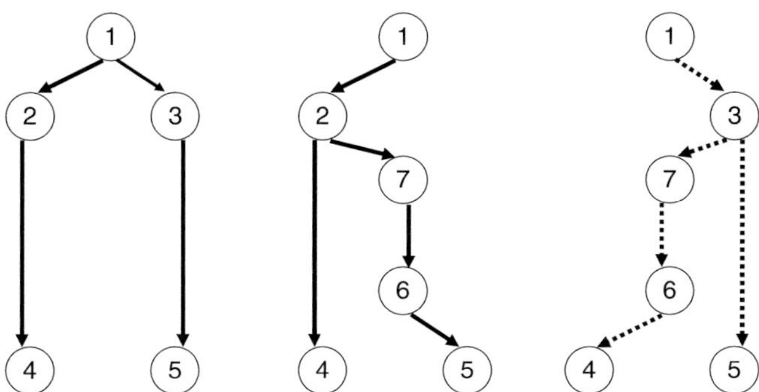

Figure 2.2 Maximum Steiner tree packing. The optimal STP for the case of unbounded delay consists of all the trees depicted above, each with weight 0.5. The case of bounded delay of one consists only of the leftmost tree with weight one

Figure 2.2 shows the optimal Steiner trees for the example in Fig. 2.1. The maximum Steiner Tree Packing problem is known to be max-SNP Hard [13, 14], both in its fractional and integer forms, i.e., there exists a constant $\epsilon > 0$, such that no polynomial algorithm exists that can approximate the optimal solution within a factor $1 + \epsilon$ of the optimum, unless $P = NP$. Nevertheless, the fractional form is considered to be easier. The best polynomial time approximation algorithm for the fractional case is able to find a solution within a factor of 1.55 of optimality [15]. The integral problem appears to be much harder. In fact, until recently, there were no polynomial time approximation algorithms to approximate the problem within any constant error factor ϵ. Very recently, however, Lau provided the best known maximum Steiner tree packing algorithm which finds an approximate solution to the problem in polynomial time which is within a factor 26 of the optimal solution [16]. The problem of efficient maximum Steiner tree packing is therefore still essentially open. Other polynomial time algorithms exist with approximation factors of $(|T| + 1)/4$, where $|T|$ is the number of receivers [13].

This chapter was a brief introduction to the problem of common information flow with and without network coding in problems with only a single source node. In the next chapter, we review a generalization of this problem when multiple sources try to each communicate with their intended set of clients.

3   Lossless multicast of multiple uncorrelated sources

In the previous chapter, we reviewed the problem of lossless multicast when there is only one information source in the network. In many practical applications, however, there is more than one information source, each with its intended set of clients. One might be interested in the maximum rate of information that can be communicated from each source to its intended clients. Unlike the case with only one information source, the problem becomes involved even when each source has only one intended client, a case called multi-unicast problem. Even then, optimum communication strategies and the set of achievable information rates are not known in general. We will review some of the known results for the multi-unicast problem.

3.1   Multi-unicast problem

We start with the multiple unicast, or multi-unicast, problem. Materials in this section are mainly a summary of the results in [17, 18]. The readers are referred to these works for further reading. The multi-unicast communication problem with k sources is defined as follows: A graph $G = \langle V, E \rangle$ and k pairs of vertices $\{(s_1; t_1); (s_2; t_2) \ldots (s_k; t_k)\}$ are given. Each source s_i is assumed to have access to an independent information source that it wishes to communicate to its intended receiver t_i. A demand vector of rates $\mathbf{r} = (r_1, r_2, \ldots, r_k)$ is said to be achievable if a communication strategy can be found to simultaneously communicate information with rate r_i from s_i to t_i for all $i = 1, 2, 3, \ldots, k$. The graph G may be either directed or undirected and the communication strategy may be based solely on routing or may also allow for network coding.

The most general problem is that of finding the set of all achievable demand vectors. A simpler version of the multi-unicast problems is the maximum uniform multi-commodity flow problem, in which one seeks to maximize the sum of the flow rates, under the constraint that all flows must be equal.

3.2   Multi-unicast with routing and network coding on directed acyclic graphs

A flow f_i of rate r_i is a collection of paths P_i from s_i to t_i and a positive real number $f_i{:}P \to \mathcal{R}^+$, such that $\sum_{P \in P_i} f_i(P) = r_i$. A k-multi-flow is a collection of k flows

$F = \{f_1, f_2, \ldots, f_k\}$ and is said to be feasible if it respects the capacity constraint on all links, i.e.:

$$\forall e \in E: \sum_{i=1,2,\ldots,k} \sum_{P \in P_i : e \in P} r_i \leq C(e).$$

The above formulation assumes routing as the only communication mechanism. When network coding is allowed, vertices in G may transcode information they receive before transmitting on an outgoing edge. For simplicity, we assume that every communication pair is demanding a rate 1 connection and each edge has capacity 1. One can transform a problem with arbitrary integer rates to a problem with more communication pairs each requesting a rate 1 connection. Similarly, by scaling the rates appropriately and adding multiple links between pairs of vertices, one can simulate arbitrary integer capacities on the edges.

3.2.1 Coding gain in directed networks can be high

By definition, the maximum multi-unicast rate with network coding is at least as large as that achievable with routing. An important question is how much improvement can one expect from network coding over routing in the multi-unicast problem. Just as in the case of single source multicast, the answer to this problem strongly depends on whether the underlying graph is directed or not.

A simple example by Harvey *et al.* [17] proves significant improvements are possible from using network coding in *directed networks*. Fix an arbitrary positive integer k and construct G on a directed graph with vertex set $V = \{s_1, s_2, \ldots, s_k, u, v, t_1, t_2, \ldots, t_k\}$. Edges are (s_i, u), (u, v), and (v, t_i) for all (s_i, t_j) and for all $j \neq i$ and $i = 1, 2, \ldots, k$. The communication pairs are (s_i, t_i) for $i = 1, 2, \ldots, k$ for all $i = 1, 2, \ldots, k$ (see Fig. 3.1). It is easy to see that with routing only, the total flow is at most 1 and thus, the maximum uniform rate of flow is $O(1/k) = O(1/n)$ where n is the number of nodes in the network. The reason is that there is only one path from s_i to t_i for each i, and that path passes through (u, v). As such, the maximum total flow is bounded by 1. With network coding, however, each s_i can send one bit to u. Then, u can send the XOR of all these k

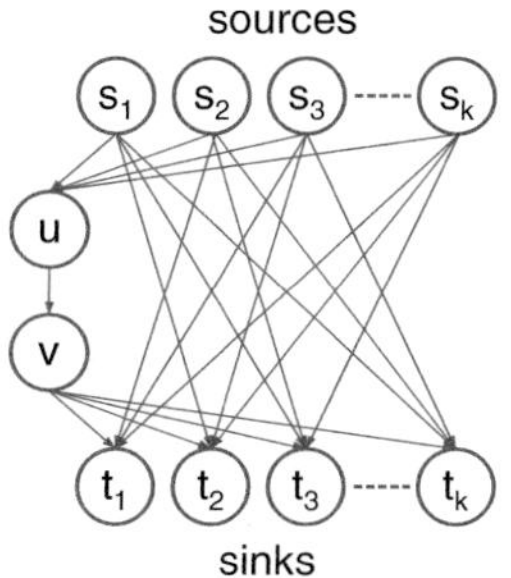

Figure 3.1 Example of a multi-unicast problem where network coding provides large gains compared to routing [17]

bits to v, where v can duplicate them to all the sink nodes. By XORing the bit received from v by all other bits directly received by $s_j, j \neq i$, the node t_i can recover the bit from s_i. Thus, a uniform throughput of 1 is simultaneously achievable by all communication pairs, a factor of k improvement over the case of routing.

The above simple example suggests a potentially high gain when network coding is used for lossless communication of multiple sources. The situation, however, appears to be different when dealing with undirected networks.

In this section we review the results on the relationship between the capacity of cuts in a graph G and the maximum concurrent information flow between all pairs of vertices in G, as derived in [17, 18].

3.2.2 Cuts in undirected graphs

In an undirected graph, a cut can be defined in two ways. First, for a subset $U \subset V$, the cut defined by U is the set of edges with one endpoint in U and one endpoint in V. Second, and more generally, a set $A \subset E$ is a cut if it separates at least one source-sink pair. In general, we can define the capacity of the cut, the demand crossing the cut, and the sparseness of the cut for either definition of a cut.

Let U be a subset of the vertices of G and let $A \subset E$ be the set of edges with one endpoint in U and one endpoint in $V - U$. The capacity $C(U, V - U) = C(A)$ of the cut defined by U is the sum of the capacities of the edges in A. The demand $D(U, V - U) = D(A)$ across the cut defined by U is equal to the total demand between source-sink pairs separated by A. The ratio of the capacity of the cut A to the demand between source-sink pairs separated by A is called the sparseness of A:

$$S(U, V - U) = S(A) = \frac{C(A)}{D(A)}.$$

For a graph G, the values of the sparsest cut is defined as:

$$S_G = \min_{A \subset E} S(A).$$

3.2.2.1 Cuts in directed graphs

For a directed graph G, Harvey *et al.* define a parameter called meagerness, whose definition is similar to the definition of sparseness but is more relevant to the problem at hand.

Given an edge set $A \subset E$ and a subset of source-sink pairs $P = f(s_i, t_i) : i \in I$, A is said to isolate P if for $i, j \in I$, every path from s_i to t_j intersects A. We denote $\sum_{i \in I} r_i$ by $D(P)$. The meagerness of the cut A, denoted by $M(A)$, is defined to be 1 if A does not separate any source-sink pair (s_i, t_i), and is otherwise defined by:

$$M(A) = \min \left\{ \frac{C(A)}{D(P)} : A \text{ isolates } P \right\}$$

where the minimum is taken over all sets of source-sink pairs. The value of the most meager cut in G is denoted by:

$$M_G = \min_{A \subset E(G)} M(A).$$

Obviously, $M_G \geq S_G$. The following theorem is true:

THEOREM 3.1   *For an undirected network, the rate of the maximum concurrent multi-unicast flow with routing is at most S_G. For a directed acyclic graph H and when network coding is allowed, the maximum rate of the concurrent multi-unicast is bounded by M_H.*

Unfortunately, these bounds can be fairly loose, as shown by the seminal work of Leighton and Rao [19].

THEOREM 3.2 (Leighton and Rao [19])   *There exists a graph G for which the rate of concurrent multi-unicast flow, achievable with routing only, is a factor $O(1/\log n)$ smaller than the sparsest cut.*

This result can be extended to the case when network coding is allowed. The following result can be proved by a simple variation of the example by Leighton and Rao in [19].

THEOREM 3.3 (Harvey *et al.* [17])   *Even when network coding is allowed, there exists an undirected graph G for which the rate of concurrent multi-unicast flow with network coding is a factor $O(1/\log n)$ smaller than the sparsest cut.*

Similarly, the meager is not a tight bound either.

THEOREM 3.4   *For the directed acyclic network in Fig. 3.2, the value of the most meager cut is at least 50% more than the maximum rate of concurrent multi-unicast flow achievable with network coding.*

For further information on the multi-unicast problem, please refer to [17, 18].

3.2.3   Coding gain in undirected networks: Li and Li's conjecture

As Fig. 3.1 shows, there exists a *directed* network in which the coding gain for the multi-unicast problem can be large. The case is very different in undirected networks. In fact, it is strongly believed that there is no coding using network coding in undirected networks, a conjecture called the Li and Li's conjecture [20]. This is to be compared to the case of multicast, where the maximum coding gain was shown to be at least 8/7 and at most 2. While the Li and Li's conjecture remains an open problem in its most general form, it has been verified on several special cases, such as bipartite and planar graphs [20–22].

3.3   Concluding remarks

In this chapter, we reviewed some of the results on multi-unicast communication in which multiple source-receiver pairs concurrently communicate over a network. Much

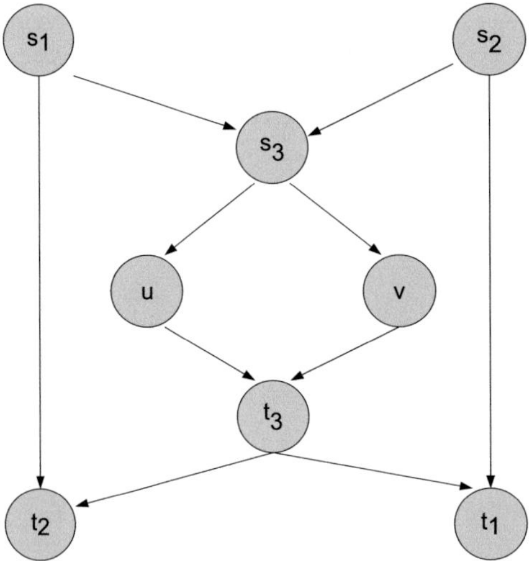

Figure 3.2 An example of a network with meager 1. For this example, Harvey *et al.* [17] show that the maximum rate of concurrent multi-unicast with network coding is at most 2/3

like the multicast scenario, network coding can provide substantial gains in directed networks. It is belived that no coding gain is possible in undirected networks. Multiple multicast problem is the natural generalization of the multi-unicast problem to the case where each source may have more than one intended client. As one could guess from the multi-unicast problem, not many concrete results exist on this general problem. In the next section, we will review a multicast problem with multiple correlated sources. Unlike the problem in this chapter, the sources may be correlated, which adds an extra dimension of difficulty to the problem. However, we will mostly focus on applications with a single set of receivers, which makes the problem trackable in many interesting scenarios.

4 Lossless multicast of multiple correlated sources

Lossless distributed source coding is a well-studied topic in source communication that has been studied for simple and general cases, with or without network coding. The Slepian-Wolf (S-W) [1] distributed source coding problem considers lossless communication of two discrete correlated sources. More precisely, let X and Y be two correlated sources on a finite alphabet. The S-W problem is to encode X and Y *separately* to R_X and R_Y bits per source symbol respectively, such that a joint decoder can recover both X and Y without error. As Slepian and Wolf proved in their seminal work [1], such encoding is possible if and only if $R_X \geq H(X|Y)$, $R_Y \geq H(Y|X)$ and $R_X + R_Y \geq H(X, Y)$, where H is the discrete entropy.

As discussed in the introduction, the S-W distributed source coding can be considered as an instance of network source coding for a special network with two source nodes (one for X and one for Y) that are directly connected to a single client. Recently, Ho *et al.* [3] generalized S-W coding to distributed source coding over arbitrary networks, with an arbitrary number of sources. We start with the simple S-W problem, before reviewing the results for S-W coding on arbitrary networks.

4.1 Slepian-Wolf problem in simple networks

Slepian-Wolf theory, also known as distributed data compression, or lossless separate coding of correlated random variables, is a typical source coding problem and belongs to the scope of network information theory. It was first studied by Slepian and Wolf in the 1970s. The diagram of this problem is shown in Fig. 4.1.

The task of Slepian-Wolf coding is to study the minimum rate needed in transmitting the sources X and Y to a common decoder losslessly in the usual Shannon sense. The theoretical result seems a little surprising at first glance. Largely speaking, it says that the minimum sum rate for the lossless compression in a distributed manner is exactly the same as in the centralized manner. Slepian-Wolf coding is very useful in data compression, especially in a distributed source coding environment such as the wireless sensor network. In the following, we will discuss, in detail, its information theoretical background and practical code design.

Let X and Y be discrete random variables with values in alphabets $\mathcal{X}$ and $\mathcal{Y}$, respectively. The joint distribution of X and Y is $p(x,y)$. Let $(X_1, Y_1), (X_2, Y_2), (X_n, Y_n)$ be a sequence of independent, identically distributed (i.i.d.) random samples generated from

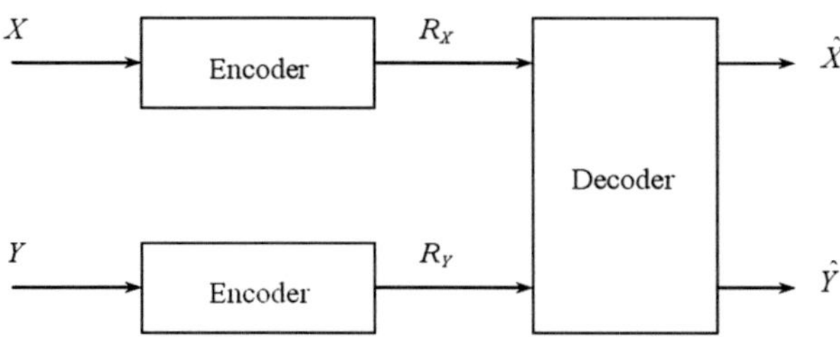

Figure 4.1 Slepian-Wolf coding

(X, Y). A distributed source code is composed of two encoders

$$f_X : \mathcal{X}^n \to 1, 2, \ldots, M_X$$
$$f_Y : \mathcal{Y}^n \to 1, 2 \ldots, M_Y,$$

and a decoder

$$g : 1, 2, \ldots, M_X \times 1, 2 \ldots, M_Y \to \mathcal{X}^n \times \mathcal{Y}^n.$$

The probability of error of a distributed source code is defined as

$$P_e^{(n)} = \Pr\left(g(f_X(X^n), f_Y(Y^n)) \neq (X^n, Y^n)\right).$$

A rate pair (R_X, R_Y) is admissible if for any $\epsilon > 0$, there exists, for sufficiently large n, a distributed source code to satisfy the following conditions:

$$\frac{1}{n} \log M_X < R_X + \epsilon$$
$$\frac{1}{n} \log M_Y < R_Y + \epsilon$$
$$P_e^{(n)} < \epsilon.$$

4.1.1   Main theorem and its proof

THEOREM 4.1 (Slepian-Wolf)   *The achievable region of the noiseless distributed source coding is given by*

$$R_X \geq H(X|Y)$$
$$R_Y \geq H(Y|X) \tag{4.1}$$
$$R_X + R_Y \geq H(XY).$$

This achievable region is shown in Fig. 4.2. Since the minimum total rate for lossless reconstruction of both X and Y equals $H(XY)$ even if the sources are jointly coded, the Slepian-Wolf coding theorem reveals a surprising fact that, in this scenario, distributed source coding can do as well as a centralized coding.

The proof of the theorem is now standard in information theory. It is composed of the achievable part, which shows that the rate pair satisfying (4.1) is always achievable and the converse part, which says that for any admissible code (4.1) must be satisfied.

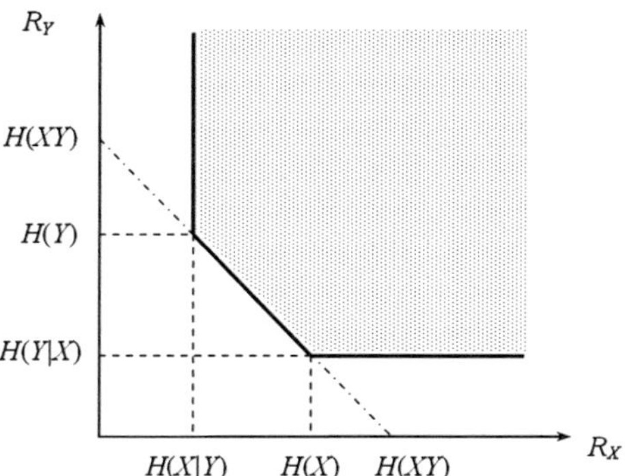

Figure 4.2 The achievable region of the Slepian-Wolf problem

Proof of the Slepian-Wolf theorem:

Achievability:

In this part, we will show that there is a coding method such that the rate pair satisfies Eq. (4.1) and the probability of error approaches zero as the block size goes to infinity. We omit some of the technicalities for ease of reading.

Random code book generation:

For each typical sequence x randomly choose (with uniform distribution) one of 2^{nR_X} bins and put the sequence into it. Thus each bin has roughly the same number of sequences. Similarly, put each typical sequence y randomly into one of 2^{nR_Y} bins. The assignment of typical sequence $x(y)$ is revealed to encoders, and both assignments are revealed to the decoder.

Encoding:

On observing the sequence x, the encoder X searches into its code book and outputs the index of the bin to which x belongs. Similarly, encoder Y outputs the index of the bin to which y belongs. If any encoder cannot find the input sequence, it declares an error.

Decoding:

On receiving both indices, the decoder searches into the two bins indicated by the indices and finds a unique pair (x, y) which is jointly typical, which is denoted by $(x, y) \in \mathcal{T}_{XY}$. If no such pair exists, or if there is more than one pair, the decoder declares an error.

Analysis of the probability of error:

The encoding-decoding procedure fails if and only if at least one of the following events occurs.

E_0: x and y are not jointly typical;

E_1: there exists $x' \neq x$, and x' and x belong to the same bin and (x', y) are jointly typical;

E_2: there exists $y' \neq y$, and y' and y belong to the same bin and (x, y') are jointly typical;

E_3: there exists x and y, $y' \neq y$, x' and x, y', and y belong to the same bin and (x', y') are jointly typical;

It is obvious that

$$P_e^{(n)} \leq \sum_{i=0}^{3} \Pr(E_i).$$

Next, we will show that for any given $\epsilon > 0$, if R_X and R_Y satisfy the condition of (4.1), then for sufficiently large n, $P_e^{(n)} < 4\epsilon$. Thus the achievable part is proved.

According to Asymptotic Equipartition Property [60], $Pr(E_0) < \epsilon$, for sufficiently large n.

$$P\left(E_1 \bigcup E_0^c\right) = \sum_{(x,y)} p(x, y) \Pr\left\{\exists x' \neq x : f_X(x') = f_X(x), (x', y) \in \mathcal{T}_{XY}\right\}$$

$$\leq \sum_{(x,y)} p(x, y) \sum_{x' \neq x, (x', y) \in \mathcal{T}_{XY}} \Pr\left\{f_X(x') = f_X(x)\right\}$$

$$= \sum_{(x,y)} p(x, y) \frac{|\mathcal{T}_{X|Y}|}{2^{nR_X}}$$

$$\leq 2^{n(H(X|Y)+\epsilon) - nR_X}.$$

Thus, if $R_X > H(X|Y)$, $P(E_1 \bigcap E_0^c) < \epsilon$ for sufficiently large n. Similarly, if $R_Y > H(Y|X)$ and $R_X + R_Y > H(XY)$, then $P(E_2 \bigcap E_0^c) < \epsilon$ and $P\left(E_3 \bigcap E_0^c\right) < \epsilon$. So $P_e^{(n)} < 4\epsilon$ for sufficiently large n. The achievable part is proved.

Converse part:

Assume for any $\epsilon > 0$ the encoding f_X, f_Y and decoding g of a distributed source code satisfy $P_e^{(n)} < \epsilon$ for sufficiently large n. Then

$$nR_X \geq H(f_X(X^n))$$

$$\overset{a}{\geq} H(f_X(X^n)|Y^n)$$

$$\overset{b}{=} H(f_X(X^n)|Y^n) - H(f_X(X^n)|X^n Y^n)$$

$$= I(X^n; f_X(X^n)|Y^n)$$

$$= H(X^n|Y^n) - H(X^n|f_X(X^n)Y^n)$$

$$\overset{c}{\geq} H(X^n|Y^n) - H(X^n|f_X(X^n)f_Y(Y^n))$$

$$\overset{d}{\geq} H(X^n|Y^n) - n\epsilon$$

$$\overset{d}{\geq} nH(X|Y) - n\epsilon,$$

where a comes from the fact that conditioning reduces entropy;

b comes from that $f_X(X^n)$ is a function of X^n;

c comes from that $f_Y(Y^n)$ is a function of Y^n and conditioning reduces entropy;

d comes from the Fano's inequality;
e comes from that (X_i, Y_i) are *i.i.d.*

Similarly, we have

$$nR_Y \geq nH(Y|X) - n\epsilon.$$

$$
\begin{aligned}
n(R_X + R_Y) &\geq H(f_X(X^n)f_Y(Y^n)) \\
&= H(f_X(X^n)f_Y(Y^n)) - H(f_X(X^n)f_Y(Y^n)|X^nY^n) \\
&= I(f_X(X^n)f_Y(Y^n); X^nY^n) \\
&= H(X^nY^n) - H(X^nY^n|f_X(X^n)f_Y(Y^n)) \\
&\geq H(X^nY^n) - n\epsilon \\
&= nH(XY) - n\epsilon.
\end{aligned}
$$

Since ϵ can be arbitrarily small, the converse part is proved.

4.1.2 Slepian-Wolf coding for many sources

The Slepian-Wolf coding can be easily generalized to more than two sources. The result is briefly summarized in the following.

THEOREM 4.2 (Slepian-Wolf for many sources)   *Given correlated sources $X_1, X_2, \ldots, X_m$. The achievable region of the noiseless distributed source coding is given by*

$$\sum_{i \in S} R_i \geq H(X_S|X_{S^c})$$

for $S \subseteq \{1, 2, \ldots, m\}$, where $X_S = \{X_i | i \in S\}$.

4.1.3 Slepian-Wolf code design

From the proof of the achievable part of Slepian-Wolf theorem, one obtains a random code design method. But this is usually unrealistic in practice. Practical Slepian-Wolf code needs a pipeline structured decoder. So, there comes the problem of how to achieve an arbitrary point on the boundary of the achievable region.

Before we come to the problem of realization of Slepian-Wolf code, we introduce some useful concepts relating to the geometry properties on the achievable region. It is obvious that this region is a *convex polyhedron*. The *dominant face* of this region is defined as all the rate pairs satisfying $R_X + R_Y = H(XY)$. *Vertices*, or *corner points* of the region, are defined as the rate pairs that occur at the intersection of bounding surfaces. In the particular case of two sources, there are two vertices and one dominant face, which is actually a line segment. The problem, in fact, reduces to achieving an arbitrary point that belongs to the dominant face. If that is possible, other points can be achieved by, say, time-sharing. Because one can use a sequential coding-decoding pair to realize any of the two vertices, time-sharing will guarantee that any point between these vertices is achievable. But the major problem with the time-sharing method is that it requires the two encoders to be synchronized. In the following, we will introduce two

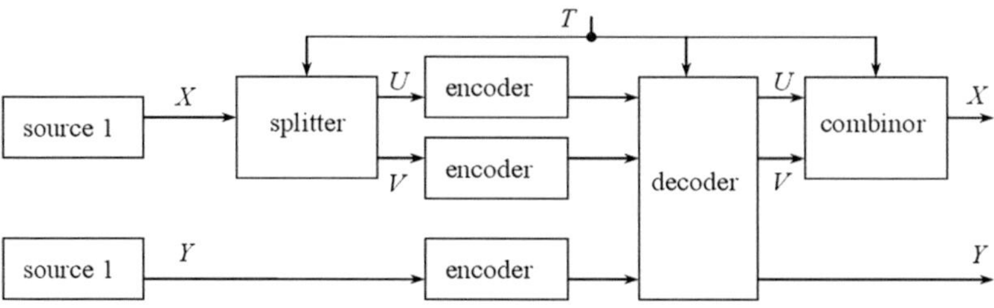

Figure 4.3 Source splitting in the Slepian-Wolf coding

methods which avoid the need for time-sharing, namely source splitting and channel code partitioning.

4.1.3.1　Source splitting

Using source splitting in Slepian-Wolf coding was introduced by Rimoldi and Urbanke [59]. In one sentence, source splitting is to make any point on the dominant face be a vertex of a Slepian-Wolf coding of higher dimension. The method of Rimoldi and Urbanke is as follows.

Assume X takes values in the alphabet $\mathcal{X} = \{1, 2, \dots, |\mathcal{X}|\}$. Set $\mathcal{U} = \mathcal{V} = \mathcal{V} \cup \{0\}$. Let T be a binary random variable taking values in $\{0, 1\}$, independent of X and Y. Let $U = X \cdot T$, $V = X \cdot (1 - T)$. $PrT = 1 = p$ is a parameter used to adjust how much information of X is contained in U. The encoders and the decoder share the same randomness T. The Slepian-Wolf coding can now be performed in three steps. First, encode U using the rate $H(U|T)$. Then with the knowledge of U, encode Y using rate $H(Y|U)$. Lastly, with the knowledge of U and Y, encode V at rate $H(V|UY)$. To verify that the sum rate equals to $H(XY)$, we have

$$
\begin{aligned}
H(XY) &= H(XYT) - H(T) \\
&= H(UVYT) - H(T) \\
&= H(UT) + H(VY|UT) - H(T) \\
&= H(U|T) + H(Y|UT) + H(V|YUT) \\
&= H(U|T) + H(Y|U) + H(V|YU),
\end{aligned}
$$

where the last step is because $U = 0$ if and only if $T = 0$. The scheme of the source-splitting method is shown in Fig. 4.3.

A property of the above source-splitting method is that it requires the encoders and decoder to share a common randomness. To avoid this, another source-splitting method was introduced by Coleman *et al.* [63]. In their settings, a splitting function is used which is defined as follows. For a discrete memoryless source $X = (X1, X2)$ taking values in an alphabet $\mathcal{X}$, define the function $f : \mathcal{X} \rightarrow \mathcal{X} \times \mathcal{X}$ as:

$$
f(X_i) = (X^a, X^b) = (\min(\pi(X), T), \max(\pi(X), T) - T),
$$

and the inverse function $f^{-1} : \mathcal{X} \times \mathcal{X} \to \mathcal{X}$:

$$f^{-1}\left(X_i^a, X_i^b\right) = \pi^{-1}\left(X_i^a + X_i^b\right),$$

where $T \in \mathcal{X}$ is a threshold and π is a permutation function of $\mathcal{X}$.

Apart from the property that the new splitting method does not require a common randomness, the alphabet of the splitting sources can also decrease. The splitting function can be generalized to more than two sources. It is proved in [64] that for $M \geq 2$ correlated sources, in order to realize the pipelined decoding strategy, splitting the sources into $2M - 1$ virtual sources is sufficient. And more specifically, at most one split per source is required. The diagram of this splitting method is similar to that in Fig. 4.3, without the common random variable T. In literature, the source-splitting method has also been applied to multi-access channels [61]. Considering the duality of multiple access channels and distributed coding of correlated sources, it is easy to understand that one method useful for one system is applicable to another.

4.1.3.2 Slepian-Wolf coding by using channel codes

Although Slepian-Wolf coding was discovered more than 30 years ago, its potential importance had not been mirrored in practical data compression for a long time. Wyner may be the first to propose to use a channel code to realize the Slepian-Wolf coding [69]. But only very recently, people began to design practical methods to realize Slepian-Wolf coding by powerful channel codes such as Turbo codes and LDPC.

Before we have a look at Wyner's idea, let's have a brief review of channel coding. Assume we have a binary additive channel with binary input X and output Y, which is determined by $Y = X + U$, where $U \in \{0, 1\}$ and is independent of X, and $+$ denotes module 2 addition. A (n, k) channel code is specified by a $(n - k) \times n$ parity check matrix $\mathbf{H}$. The code $\mathcal{C}$ consists of all n-sequence x (which is a column vector) such that its syndrome, which is defined by $s = \mathbf{H}x$, is zero. The set of all n-sequences having the same syndrome is called a coset, usually denoted by $\mathcal{C}_s$. On receiving y, the decoder calculates its syndrome $s = \mathbf{H}y$. The decoding function $f(s)$, which is a map from $(n - k)$-sequences to n-sequences, is the codeword v in the coset $\mathcal{C}_s$ which has the minimum Hamming weight. The n-sequence v is also called the coset leader of the coset $\mathcal{C}_s$. The channel decoder declares that the input n-sequence should be $\hat{x} = y + v$.

Because U is independent of X, so the virtual channel connecting X and Y can be considered as a binary symmetric channel (BSC) with crossover probability p, which in short is denoted by BSC(p). We have the following proposition.

PROPOSITION 4.1 *For any $\epsilon > 0$, for sufficiently large n, there exists a (n, k) parity-check code with $k/n > 1 - H(p)$, where $1 - H(p)$ is the channel capacity of BSC (p), such that the probability of error is no greater than ϵ.*

Remark: Generally, we do not have a counterpart of the above proposition for discrete channels other than BSC. So, that's why the following result is only used in binary cases.

In Wyner's scheme, the encoder observes Y, calculates its syndrome $z = \mathbf{H}y$, then transmits z to the decoder. The decoder calculates $z + \mathbf{H}x = \mathbf{H}(x + y) = \mathbf{H}u$. Then let $\hat{y} = x + f(\mathbf{H}v)$. From the above proposition, one can see that the source Y can be recovered with arbitrarily small error. This procedure is usually called the "syndrome decoding." As a special case of the Wyner's idea, if X vanishes, one can still use a channel coding method to realize the near-lossless source coding, which traditionally is performed by entropy codes.

In Wyner's scheme, in order to realize any point on the boundary of the Slepian-Wolf coding, one has to achieve the vertices first. After that, the time-sharing argument is employed. For any vertex, say $(H(X), H(Y|X))$, one can use one entropy code (to achieve $H(X)$) plus one parity-check code with syndrome decoding (to achieve $H(Y|X)$), or two parity-check codes with syndrome decoding.

Since, as was demonstrated in the previous section, time-sharing requires synchronized encoders, researchers want to find some methods which can realize any point on the dominant face of the achievable region. In [55], a distributed code construction for the entire Slepian-Wolf rate region is proposed. We introduce their idea with the assumption that X and Y are connected by a virtual BSC, $Y = X + U$, where $U \in \{0, 1\}$ is the channel noise. First, construct a generator matrix by stacking three sub-matrices whose dimensions are $n(1 - H(Y)) \times n$, $n(H(Y) - H(X)) \times n$, and $n(H(X) - H(X|Y)) \times n$, respectively. These sub-matrices are denoted by $\mathbf{G}_y$, $\mathbf{G}_a$, and $\mathbf{G}_s$.

The two encoders transmit the syndromes of the n-sequence x and y using $\mathbf{H}x$ and $\mathbf{H}y$, respectively. At the decoder side, the syndromes of x and y are such that the following equalities hold:

$$x = u_{cx}\mathbf{G}_c + u_{ax}\mathbf{G}_a + u_{sx}\mathbf{G}_{sx} + s_x$$
$$y = u_{cy}\mathbf{G}_c + u_{sy}\mathbf{G}_{sy} + s_y.$$

Denoting $u = [u_{cx} + u_{cy}|u_{ax}|u_{sx}u_{sy}]$ and $s = s_x + s_y$. It can be seen that $x + s_x$ is a valid codeword of C_x, which is formed by the generate matrix $\mathbf{G}x$. Because C_x is a subcode of C, $x + s_x$ is a valid codeword of C. Similarly, $y + s_y$ is a valid codeword of C and so is $x + y + s$. The task of the decoder is then to find a codeword that is closest to s and announce it to be $x + y + s$. The following proposition gives a sufficient condition that this decoder has zero error.

PROPOSITION 4.2   *For given generator matrices $\mathbf{G}$, if* weight $(x + y) \le t$, *where t satisfies $d_{min} \ge 2t + 1$, and d_{min} is the minimum distance of the code generated by $\mathbf{G}$, the decoder has zero error.*

The proof of the proposition is quite clear. The decoder fails if and only if for given s, there exists another source pair x' and y' such that $x' + y' + s$ is a codeword and $weight(x' + y') \le t$. But then the distance between the two codewords $x + y + s$ and $x' + y' + s$ will be no greater than $2t$, which is a contradiction to the assumption $d_{min} \ge 2t + 1$.

After the correct decoding of $x + y$, the decoder calculates u according to

$$x + y = u\mathbf{G} + s.$$

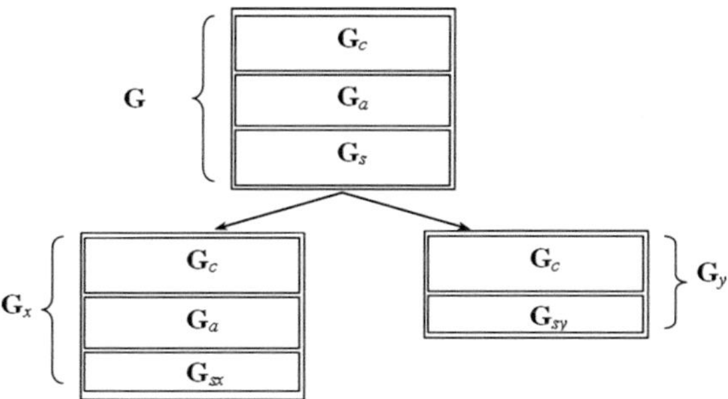

Figure 4.4 Splitting of the generate matrix **G**

For example, if **G** is in its systematic form, $\boldsymbol{u}$ is composed of the first $nH(X|Y)$ elements of the codeword. Since $\boldsymbol{u}_{sy}$ is also known, one can use it to decode $\boldsymbol{y}$ and then $\boldsymbol{x}$ can be decoded.

Two special cases of the splitting correspond to the corner points of the Slepian-Wolf region. When $\mathbf{G}_{sx} = $ null, the rates are $H(X)$ and $H(Y|X)$ respectively. And when $\mathbf{G}_{sy} = $ null, the two rates are $H(Y)$ and $H(Y|X)$ respectively. Similar ideas were also used in [65] and [66].

In [68], the authors use LDPC to achieve the boundary of the Slepian-Wolf region of correlated binary symmetric sources. The idea is that, instead of transmitting syndromes, the encoders use parity checks, as discussed below.

Assume we want to realize any rate pair (R_X, R_Y) satisfying $R_X + R_Y = H(X, Y)$. An LDPC code with rate $k/n = 1/(1 + H(p))$ is used, where p is the crossover probability of the virtual BSC channel connecting X and Y. For the X encoder, for the input k-sequence, use the LDPC code to generate $n - k$ parity-check bits. Then output the first $1/a$ fraction of the information bits and the first $1 - 1/a$ fraction of the parity-check bit. The Y encoder uses the same LDPC code, and outputs the remaining $1 - 1/a$ fraction of the information bits and the remaining $1/a$ fraction of the parity-check bits (Fig. 4.5).

At the decoder side, in order to decode X, three parts of bits are used: the $1/a$ fraction of the information bits of X (which is error free), $1 - 1/a$ fraction of the information bits of Y (which is the output of a virtual BSC channel with the input X), and the $1 - 1/a$ fraction of the parity-check bits (with the remaining $1/a$ fraction considered as punctured) of X. Y is decoded similarly. One can verify that if we choose an LDPC with rate $k/n = 1/(1 + H(p))$, the Slepian-Wolf coding rates are:

$$R_X = [k/a + (n - k)(1 - 1/a)]\,k = 1/a + (1 - 1/a)H(p)$$
$$R_Y = [k(1 - 1/a) + (n - k)/a]\,k = 1 - 1/a + H(p)/a.$$

So, $R_X + R_Y = 1 + H(p) = H(XY)$. By adjusting $a \geq 1$, one can achieve any point on the dominant face of the Slepian-Wolf region. For the symmetric case, i.e., the channels require the same rate, one can use an LDPC code with rate and without puncture.

The practical realization of the Slepian-Wolf coding relies on the parity-check codes used. Until now, it is only clear that for a BSC, there exists a good parity-check code that

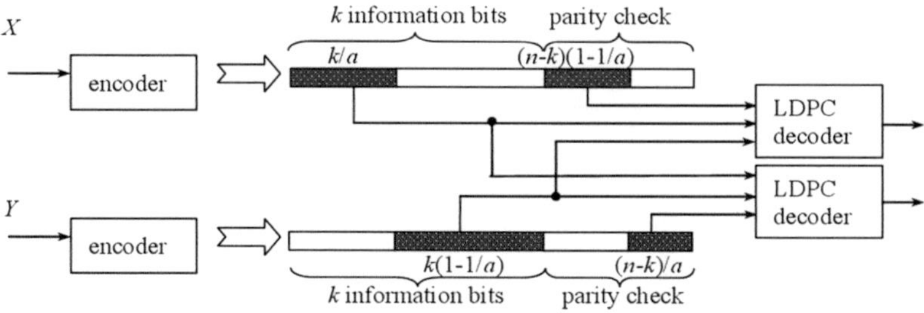

Figure 4.5 Slepian-Wolf coding using parity checks

can achieve the channel capacity. Considering the duality of source coding and channel coding, a good channel code can be used as a good source code. In [56], it is directly proved that a Mackey structured LDPC code can be used as a good source code for a binary symmetric source (BSS).

In summary, there are two methods used in the channel code based Slepian-Wolf coding: the syndrome decoding method and the parity-check bits method. When using the syndrome decoding method, one can always employ a traditional channel code such as LDPC and perform the usual decoding procedure. But when using the parity-check method, in order to achieve the whole rate region, one needs to use a channel decoding method facing parallel channels. That is, one needs to consider the fact that different bits of the virtually received sequence suffer from different kinds of noise. For example, when the real channel is lossless, the parity-check bits are error free. If the real channel is noiseless, syndrome decoding and parity approach are equivalent. In fact, there is a one-to-one correspondence between optimal syndrome approach and optimal parity-check approach. But if the real channel is noisy, parity approach will have the advantage of more consistent performance and is more suitable for a practical system [105].

4.2 Slepian-Wolf problem in general networks

So far in this chapter, we have studied the theoretical and practical aspects of the classic Slepian-Wolf problem. Here, the nodes encoding the correlated sources directly communicated to the sink node, whose job was to losslessly recover these correlated sources. This could be regarded as a network source coding problem with a simple, star topology. The same problem can be generalized to arbitrary networks, i.e., the correlated sources may reside on nodes of a network with arbitrary topologies. Unlike the simple case studies so far, the details of how source encoding is communicated through the network become important.

In general, one could use network coding to effectively communicate these encodings. The problem of network coding for correlated sources, or the Slepian-Wolf problem for general networks, was studied by Ho *et al.* in [103] where the authors characterize the achievable rate region for arbitrary networks. They also demonstrate that random linear coding can achieve the performance bounds. On one hand, these results can be considered the generalization of Theorem 4.2 for Slepian-Wolf coding for many sources to the case in which sources have to be communicated over an arbitrary network. On the other hand, these results can be regarded as the generalization of the network coding theorem to the case of compressible, correlated sources. In this section, we review these results. We refer the reader to [103] for more details.

Let $S = \{s_1, s_2, \ldots, s_{N_s}\}$ be the set of all source nodes. Each source node s_i observes a discrete, i.i.d., random variable X_i, for $i = 1, 2, \ldots, N_s$. The sources $X_1, X_2, \ldots, X_{N_s}$ may be correlated. Let $T = \{t_1, t_2, \ldots, t_{N_t}\}$ be the set of the receivers. The goal is to reconstruct *all sources losslessly at all the sink nodes*. If such lossless reconstruction is possible, we call the corresponding instance of the Slepian-Wolf problem feasible.

Let $f(s_i, t_j)$ for $i = 1, 2, \ldots, N_s$ and $j = 1, 2, \ldots, N_t$, be the max-flow from source node s_i to sink node t_j. For each sink t_j, define the rate region C_{t_j} as follows:

$$C_{t_j} = \left\{ (R_1, R_2, \ldots, R_{N_s}) : \forall B \subset S, \sum_{i \in B} R_i \leq min - cut(B, t_j) \right\}.$$

It is easy to see that if $(R_1, R_2, \ldots, R_{N_s}) \in C_{t_j}$, then a source of rate R_i can simultaneously be communicated from source node s_i to sink t_j using routing only.

Now let R_{SW} be the Slepian-Wolf region for many sources in Theorem 4.2, i.e.:

$$R_{SW} = \left\{ (R_1, R_2, \ldots, R_{N_s}) : \forall B \subset S, \sum_{i \in B} R_i \geq H(X_B | X_{B^c}) \right\},$$

where X_B, for $B \subset S$, is the set of all sources that belong to the nodes in B and $B^c = S - Y$, and $H(\cdot|\cdot)$ is the conditional entropy between two sets. The following theorem is proved in [103].

THEOREM 4.3 *An instance of the Slepian-Wolf problem on a given network is feasible, if and only if $R_{SW} \cap C_{t_j} \neq \phi$ for all $j = 1, 2, \ldots, N_t$. Moreover, if the feasibility condition is met, random linear codes are sufficient to recover all sources losslessly at all sink nodes.*

Apart from characterizing the feasible region of the problem, the work of Ho *et al.* introduces a new concept in source and network communication, namely that of *joint network source coding*. As discussed in this chapter, the Slepian-Wolf region can be achieved using linear source codes only. As discussed in Chapter 2, the single source network coding feasible region can be achieved using random linear codes, as well. The linear code used by Ho *et al.* to communicate and recover the sources at the sinks is, in effect, both a source code and a network code, or a joint network source code.

A distributed source coding strategy might be interested in separating distributed source coding from network coding. In that case, each source is compressed individually and is then multicast, using network coding, to all receivers. The question is whether or not this strategy is always feasible. More precisely, suppose that conditions in Theorem 4.3 are satisfied, i.e., for a given network, we know that there exists an encoding that can communicate a set of correlated sources $X_1, X_2, \ldots, X_{N_s}$ losslessly to a set of clients $t_1, t_2, \ldots, t_{N_t}$. The question is whether or not a separate source and network coding strategy also exists that can losslessly communicate all sources to all sinks.

This problem has been considered by Ramamoorthy *et al.* in [4], where it was shown that separation is always feasible if, (a) either there is only a single source or a single sink, in which case the problem reduces to the usual network coding; or (b) there are no more than two sources and no more than two sinks. For other scenarios, they find counter-examples for which separation does not hold. The readers are referred to [4] for more details.

4.3   Concluding remarks

The Slepian-Wolf system has many extensions. For a more general class of multi-terminal source coding systems, readers are referred to [101], in which a unified treatment is introduced. The Slepian-Wolf coding theorem also has its applications and extensions in network coding. In [140], the Slepian-Wolf rate region is used as a constraint of the optimization problem of rate allocation for given transmission structures. In [100], the Slepian-Wolf coding theorem is also used in the reliable communication through a sensor reachback system. Since Slepian-Wolf coding is optimal from the source coding point of view, whether it is still optimal in a network scenario aroused many investigations. For the special case of two sources and two receivers, the Slepian-Wolf coding is still optimal as part of a source-network separable coding. But generally this is not true. Readers are referred to [102], [104] and the references therein.

This concludes our treatment of lossless source communication in networks. We reviewed network coding ideas, and how they may improve the throughput in applications with multiple sources and receivers. In the second part of this book, we will study the problem of lossy network communication of sources.

Part II

The lossy scenario

5 Lossy source communication: an approach based on multiple-description codes

In Part I of this book, we investigated the problem of lossless source communication in networks. We reviewed the current literature, with an emphasis on new results in network coding. This chapter is the beginning of Part II, which deals with the case of lossy source coding and communication, where we allow imperfect reconstruction of sources at receivers.

There are two cases where such lossy extension to network information flow problems are required/necessary. For one, most multimedia signals, such as audio, video, and images, simply cannot be encoded losslessly. The digitization process is, by nature, lossy. With multimedia signals claiming the largest share of the traffic in today's Internet, the study of lossy network communication is particularly important. Second, lossless communication implies that the same information content has to be delivered to all receivers (the so-called common information multicast). For most multimedia applications, different classes of receivers with different bandwidth resources are often interested in consuming the same multimedia content. In such heterogeneous network environments, one might consider communicating different encodings of the same content to different receivers, such that the receivers with higher bandwidths are able to reconstruct the signal at a higher quality. The concept of quality, and the tradeoff between the rate and distortion, are the defining characteristics of lossy network communication.

In this chapter, we introduce a number of techniques that allow one to optimize the quality of delivered multimedia signals to receivers over a heterogenous network application. Some of these techniques are based on a special form of source coding, called multiple-description coding (MDC), and use routing for network delivery. In other chapters, we will review other techniques based on other forms of source coding, as well as other forms of network delivery, such as network coding.

5.1 Separating source coding from network multicast

The multicast network information flow problem (formulated earlier in Chapter 2) is concerned with the maximum rate h with which *common* information can be communicated to sink nodes of a network. For the case of a single source, this maximum rate is upper bounded by the smallest of the max-flows into all the sink nodes. The beauty of Theorem 2.1 is that it proves this common information rate to be achievable by network coding.

Now let us try to use this method to communicate a source signal X with distortion-rate function $D_X(\cdot)$ from a server node that observes and encodes X to a number of sink nodes. Equipped with network coding tools and determined to separate the source coding from network coding, one can encode the source optimally and then try to communicate the encoding to the sink nodes by multicast using network coding. We call this approach *separate source and network coding*. Let $(a_t; t \in T)$ be the vector of max-flows into sink nodes. Then, by separating source coding from network coding, the source can be communicated from a server node s to the sinks in T with distortion:

$$d_S = D_X \left(\min_{t \in T} a_t \right)$$

In other words, the quality of the reconstruction at all sink nodes is bounded by the quality of reconstruction at the node with the smallest max-flow, thus starving the nodes with higher flow capacities. However, when the max-flow into all sink nodes is equal, i.e., $a_t = h$ for all $t \in T$, separate source and network coding is optimal, as is stated in the following theorem.

THEOREM 5.1   *If the max-flow into all sink nodes is equal to h, separate source and network coding is an optimal solution to the NASCC problem.*

Proof   From Theorem 2.1, a common information of rate h bits per source symbol can be communicated to all sink nodes $t \in T$. Now encode the source optimally into h bits per source symbol and communicate it to all sink nodes. Therefore, the source can be reconstructed with a distortion $d_t = D_X(h)$ for all $t \in T$. Since the max-flow into each sink node is h, the maximum mutual information between the server node and any sink node is at most h bits per source symbol. With the definition of rate-distortion function, the minimum possible distortion in reconstructing the source signal is $d_t \geq D_X(h)$ for all $t \in T$, which proves the optimality of separate source and network coding. $\square$

Thus, when flow properties of sink nodes are homogeneous, source coding and network coding are separable. In other cases, simple examples (some of which are reported in Chapter 5) can be constructed for which separate source and network coding is suboptimal.

Non-separability of source coding from channel coding is a common theme of most classic multi-terminal information theory problems. In almost all scenarios of communication of source signals over noisy broadcast channels, source coding can not be separated from broadcast channel coding unless the communication channel to each of the receivers has the same marginal capacity (see e.g., [23]). This is analogous to Theorem 5.1. If the max-flow to individual sink nodes is the same, then the source and network coding are separable. For other scenarios, we study algorithmic solutions to the problem of signal communication in networks, or NASCC, as referred to in this book, that go beyond separate source coding and network communication.

We start by introducing a systematic approach for constructing an optimized solution to the NASCC problem with bounded delay. The main elements of this approach are multiple-description coding (MDC) to produce multiple code streams from the source signal, and a new form of diversity routing, called the Rainbow Network Flow (RNF), to optimally route these coded streams to sink nodes within bounded delays.

This chapter is organized as follows. In Section 5.3, we show that MDC provides a powerful means for NASCC by exploiting different diversity paths in the network. This, however, is only possible with optimized routing and duplication of an MDC stream at the relay nodes. We consider RNF, the problem of optimized routing of MDC streams, in Section 5.4. RNF is different from classic maximum network flow problems in that, unlike the flow of commodities, a data packet can be duplicated in the intermediate nodes. On the other hand, duplicates of one data packet received at a node are not going to be beneficial for that node. The problem is therefore to design the routing and duplication strategies to maximize the total *fresh* information received over all sink nodes.

One practical solution to the NASCC problem involves joint optimization of MDC design and routing. Finding the exact solution to this joint problem is hard. We examine a systematic approach which involves solving the RNF problem separately and then optimally designing the MDC for the optimized flows. For a certain class of MDC, the latter problem becomes a straightforward convex optimization problem with linear constraints.

5.2 Beyond common information multicast: rainbow network flow

In the multicast settings discussed above, the goal was to communicate common information to all the nodes involved. By packing Steiner trees, one ensures that information communicated to all sink nodes is the same. As will become clear in this work, to multicast a multimedia signal, we do not need to confine ourselves to common information multicast.

Consider the example in Fig. 5.1. In this example, node 1 is the source and all other nodes (2–5) are sink nodes (a broadcast problem). When the goal is to broadcast common information, the communication rate will be bounded by the max-flow into the weakest nodes (nodes 2, 3). The corresponding Steiner tree is depicted in part (b). Now consider the flow of two streams indicated in part (c). This flow ensures that both streams are

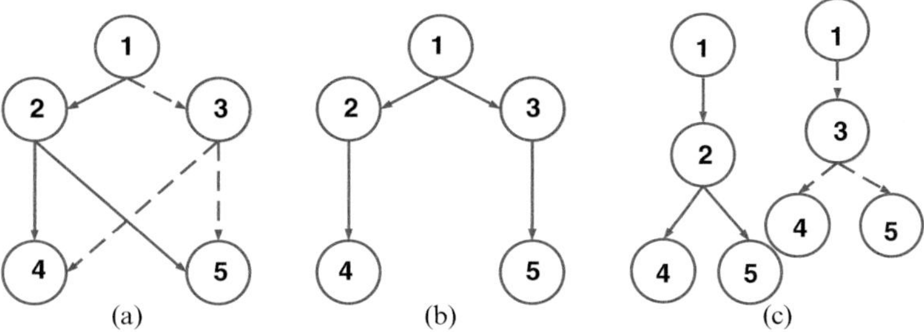

Figure 5.1 (a) Example of a network with server node 1 and sink nodes 2–5. The capacity of all links is C. (b) The best Steiner tree to broadcast common information. (c) Two distribution trees, each of which can distribute different information with rate C. While the information received by nodes 4, 5 is the same, the two nodes 2, 3 will receive different information bits

available to the nodes 4, 5 while a single stream is available to nodes 2, 3. However, the two streams available to nodes 2, 3 are different. It is easy to see that this observation holds regardless of the delay constraint. In other words, if after n time steps we insist on nodes 4, 5 to receive $2nC$ bits of information; the nC bits of information available to nodes 2, 3 have to be distinct.

As another example, consider the network in Fig. 2.1 again with a delay constraint of one. Figure 2.1(b) shows a routing that can deliver both streams to node 5 and a single stream to node 4, while still meeting the delay constraint. Note that in the classic sense of information multicast, this is not a valid multicast distribution. Unlike the example in Fig. 5.1, however, when the delay constraint is removed, both nodes can receive a larger common information at an average rate of $1.5C$ each.

In contrast to the classic, common information multicast, with a proper form of source coding, we can exploit the available data at the sink nodes to reconstruct an approximation of the source signal encoded at server nodes. In particular, one could use multiple description technique to encode the source signal into independently decodable streams. The sink nodes will be able to reconstruct the source given any subset of these streams. *Not all the streams have to be available at a sink node*; rather, the more streams that reach a sink node, the better the reconstruction quality that sink node can have.

When the data streams received by all nodes are not required to be the same, a question arises: which subset of data streams should be received by a given node v? The question becomes even more important when there is more than one server node that can provide the same data stream. There are two relevant observations from an optimization point of view. First, a node is not interested in receiving the same data stream from more than one server node. Second, the reconstruction quality at a given node, in general, depends on the specific subset of the source code streams received by the node. The dependence of the reconstruction on the specific subset of MDC streams is relaxed for a specific class of MDC, called the "balanced codes" in which the reconstruction depends only on the number of streams available for decoding. We will discuss the use of this class of MDC in detail later in this Chapter 5.

5.3   Multiple-description coding: a tool for NASCC

Multiple-description codes (MDC) have always been associated with robust networked communications, because they are designed to exploit the path and server diversities of a network. The present active research on MDC is driven by growing demands for real-time multimedia communications over packet-switched lossy networks, like the Internet. With MDC, a source signal is encoded into a number of code streams called descriptions, and transmitted from one or more server nodes to one or more destinations in a network. An approximation to the source can be reconstructed from any subset of these descriptions. If some of the descriptions are lost, the source can still be approximated by those received. This is why there seems to be a form of consensus in the literature that multiple-description codes should only be used in applications

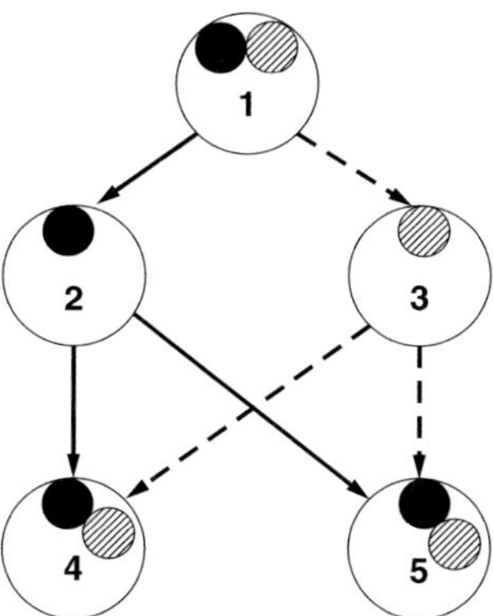

Figure 5.2 An example of flow of a two-description code

involving packet loss, because only in this case can the overhead in the communication volume be justified.

In this chapter, we show that MDC is beneficial for lossy communication even in networks where all communication links are error free with no packet loss. In this case, multiple-description coding, aided by optimized routing, can improve the overall rate-distortion performance by exploiting various paths to different nodes in the network. This is best illustrated through some examples.

5.3.1   Example 5.1

In Fig. 5.2, a server node (node 1) feeds a coded source into a network of four sink nodes (nodes 2–5). The goal is to have the best reconstruction of the source at each of these four nodes. All link capacities are C bits per source symbol. MDC encodes the source into two descriptions (shown by solid and dashed small circles in the figure), each of rate C. Descriptions 1 and 2 are sent to nodes 2 and 3 respectively. Node 2, in turn, sends a copy of description 1 to nodes 4 and 5, while node 3 also sends a copy of description 2 to nodes 4 and 5. In the end, nodes 4 and 5 will each receive both descriptions, while nodes 2 and 3 will only receive one description. For this example, increasing the delay does not increase the possible communication volume to either of the nodes beyond the one depicted in Fig. 5.2.

To see how the nodes in the network benefit from MDC, let $D_1(C), D_2(C), D_{12}(C)$ be the distortion in reconstructing the source given description 1 or 2 or both. Let $\mathbf{d} = (d_2, d_3, d_4, d_5)$ be the vector of the average distortions in reconstructing the source at nodes 2 through 5. Therefore,

$$\mathbf{d} = (D_1(C), D_2(C), D_{12}(C), D_{12}(C)).$$

For the problem at hand, let's define $\mathcal{D}_M$ as the set of all achievable distortion 4-tuples $\mathbf{d}$. Although this approach is in general a special form of NASCC, $\mathcal{D}_M$ still contains a large and interesting subset of all achievable distortion tuples. As will be shown later in this section, for this example, $\mathcal{D}_M$ includes the distortion region achievable by separate source and network coding.

The inefficiency of separate source and network coding lies in that even though nodes $4, 5$ have twice the incoming capacity compared to nodes $2, 3$, their reconstruction error ($d_4 = d_5$) is bounded by the reconstruction error of the weaker nodes ($d_2 = d_3$). Unlike lossless coding, lossy codes can play a tradeoff between the reconstruction errors at different nodes, generating a much larger set of achievable distortion tuples $\mathcal{D}_M$ than that achievable through separate source and network coding.

Tradeoffs between the quality of reconstruction at different nodes are essential in practice. For instance, in networked multimedia applications over the Internet, where the network consists of a set of heterogeneous nodes, the experience of a user with broadband connection should not be bounded by that of a user with a lesser bandwidth. Such tradeoffs are perhaps best treated as an optimization problem by introducing appropriate Lagrangian multipliers (or weighting functions). An objective function to minimize can be defined as

$$\overline{d}(\mathbf{p}, \mathbf{d}) = \mathbf{p} \cdot \mathbf{d}^T \tag{5.1}$$

where $\mathbf{p} = (p_2, p_3, p_4, p_5)$ is an appropriate weighting vector. An optimal solution corresponds to a distortion vector:

$$\mathbf{d}^*(\mathbf{p}) = \arg \min_{\mathbf{d} \in \mathcal{D}_M} \overline{d}(\mathbf{p}, \mathbf{d}).$$

Once the optimal distortion vector $\mathbf{d}^*$ is found, one should, in principle, be able to find a multiple-description code that provides the marginal and joint distortions corresponding to $\mathbf{d}^*(\mathbf{p})$ (such an MDC exists, since $\mathbf{d}^*$ has been chosen from $\mathcal{D}_M$).

As a concrete example, let's optimize the average distortion at all nodes 2 through 5 in Fig. 5.2 for $\mathbf{p} = (1/4, 1/4, 1/4, 1/4)$, in which case:

$$\overline{d} = \frac{2D_{12}(C) + D_1(C) + D_2(C)}{4}. \tag{5.2}$$

To be specific, assume that the source in question is an i.i.d. Gaussian with variance one. In this case, achievable distortions in multiple-description coding are completely derived by Ozarow in [2]. Ozarow proves the following region to be achievable when the rate of each description is C:

$$\begin{aligned}
D_1 &\geq 2^{-2C} \\
D_2 &\geq 2^{-2C} \\
\end{aligned} \tag{5.3}$$

$$D_{12} \geq \frac{2^{-4C}}{1 - \left(\sqrt{(1 - D_1)(1 - D_2)} - \sqrt{D_1 D_2 - 2^{-4C}}\right)^2}.$$

The symmetry in indices 1 and 2 ensures that (5.2) is minimized under the constraint in (5.3) when the two descriptions are balanced, that is, $D_1(C) = D_2(C) = D$. Ozarow's result, when specialized to balanced MDC states that the following set of distortions is achievable:

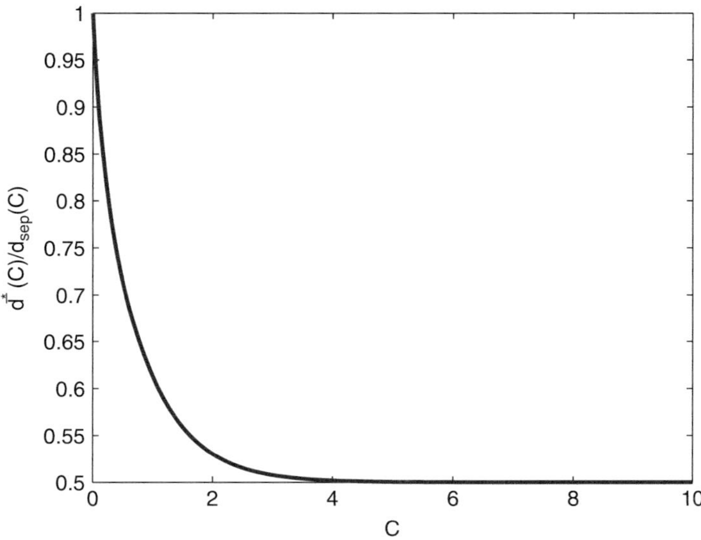

Figure 5.3 The ratio of minimum average distortion obtained by an optimized MDC over the one obtained by separate source and network coding, for the example in Fig. 5.2

$$D_1 = D_2 = D \geq 2^{-2C}$$

$$D_{12} \geq \frac{2^{-4C}}{(D + \sqrt{D^2 - 2^{-4C}})(2 - D - \sqrt{D^2 - 2^{-4C}})}. \tag{5.4}$$

The average distortion $\overline{d}$ in (5.2) can therefore be minimized under the constraints of (5.4). This is a particularly easy task because the region (5.4) is convex. Let this optimal average distortion be $\overline{d}^*_M(C)$. By separating source from network coding, the reconstruction distortion at nodes 2 through 5 (and hence the average distortion over all these nodes) is at best $d_S(C) = 2^{-2C}$. It is easy to show that $\overline{d}^*_M(C) < d_S(C)$ for all $C > 0$ (see Fig. 5.3). In other words, for all $C > 0$, there exists a balanced two-description code for which the average distortion over all sink nodes is strictly less than the average distortion achievable by any separate source and network coding scheme.

5.3.2 Example 5.2

Now consider a more involved example depicted in Fig. 5.4. A source at node s is to be communicated to nodes 1 to 8. The goal again is to minimize the average distortion over all these 8 sink nodes. All links have capacity C bits per source symbol. We choose to use an MDC with 3 descriptions, each of rate C. These descriptions are denoted by three patterns. Figure 5.4 shows a routing strategy that delivers these descriptions optimally to all the 8 nodes. The routing is optimal since each node receives a number of distinct descriptions exactly equal to its incoming capacity.

Once the routing is optimized, one still needs to optimize the MDC. Unlike the case of two descriptions, there is no closed-form representation of the rate-distortion behavior of a balanced 3-description code even in the case of a Gaussian source. Therefore,

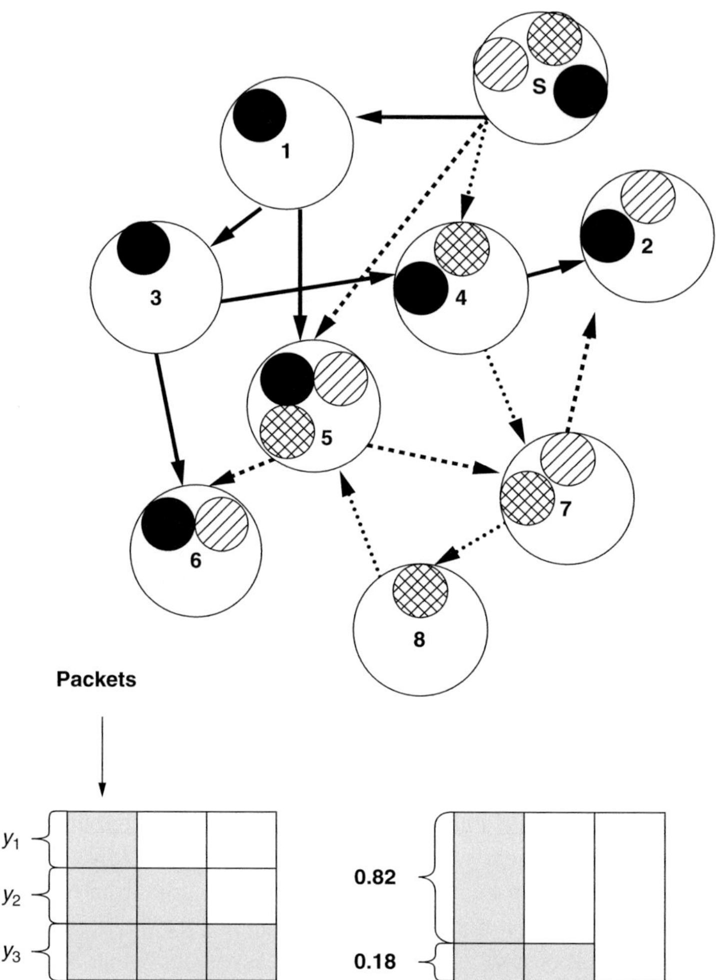

Figure 5.4 An example of optimal flow of a 3-description code: The bottom of the figure depicts the procedure of designing a 3-description MDC using Priority Encoding Transmission technique, as explained in Section 5.5

we resort to a practical technique of Priority Encoding Transmission (PET) [24] for producing whatever required number of balanced descriptions.

In particular, as is shown in Section 5.5, for any progressively refinable source with distortion-rate function $D(R)$, an MDC of K descriptions each of rate r can be constructed using PET such that, given any $k \leq K$ of the descriptions, the source can be reconstructed with distortion at most $D\left(r\sum_{l=1}^{k} ly_l\right)$, where $\mathbf{y} = (y_l; l = 1, 2, \ldots, K)$ is any positive vector such that $\sum_{l=1}^{K} y_l = 1$. There is a one-to-one correspondence between an MDC constructed by PET and any such vector $\mathbf{y}$.

It should be noted that the set of possible balanced MDCs is strictly larger than the one achieved through PET, as has been recently shown in [25]. PET, however, provides a convenient parameterization of the rate-distortion behavior of the resulting MDC. Furthermore, once a multimedia source is encoded into a progressive source code stream, creating the corresponding MDC streams through PET is straightforward. This makes PET a compelling solution given that progressive source coding is already part of existing and emerging multimedia standards such as JPEG2000 [26] for images and Fine Granularity Scalability provisions for video in MPEG4 [27].

Optimizing the MDC within the class of PET codes will result in a convenient convex optimization problem. For the example in Fig. 5.4, three, four, and one nodes receive 1,2,3 distinct descriptions respectively. For a PET code with vector $\mathbf{y} = (y_1, y_2, y_3)$, the average distortion at the sinks, assuming a Gaussian source of variance one, can be written as:

$$\overline{d} = 8^{-1} \left(3 \cdot 2^{-2Cy_1} + 4 \cdot 2^{-2C(y_1+2y_2)} + 2^{-2C(y_1+2y_2+3y_3)} \right).$$

Minimizing the above over all $y_1, y_2, y_3 \geq 0$ such that $y_1 + y_2 + y_3 = 1$ is a convex optimization problem with linear constraints. For $C = 1$, the optimal solution is $y_1 = 0.82, y_2 = 0.18, y_3 = 0$.

5.3.3   Design issues

The preceding examples expose a number of design issues, which are the subject of the rest of this chapter and Chapter 6.

- To maximize the benefit of MDC in the network, the flow of MDC streams needs to be optimized. This optimization involves appropriate routing and duplication of MDC streams at the relay nodes. Each node has to decide which received descriptions to duplicate and communicate over which outgoing edges.
- In the above two examples, optimizing the routing of MDC descriptions and optimizing the MDC design happen to be separable. This is because in both examples, each node with incoming capacity nC receives exactly n distinct MDC streams of rate C and is hence fully satisfied in flow volume. Therefore, if MDC streams of equal rate C were going to be used, the optimization of the flows could be separated from the optimization of the MDC.

  The two optimizations should be carried out jointly in general. This, however, will leave the problem intractable for most network topologies. Therefore, our strategy is to carry out the optimizations separately.
- Unlike the two examples reported in this section, optimizing MDC may result in a different number of potentially unbalanced descriptions. In this chapter, we confine ourselves solely to the case of balanced MDCs.
- The *total number of descriptions and their rates* should also be chosen appropriately. Given that we will use balanced MDCs, it is shown in Chapter 7 that using more descriptions of "lower" rate will not decrease the set of possible achievable distortions. In Chapter 7, we consider the role of description rates and the total number of descriptions formally, and characterize the set of all possible solutions achievable by our proposed method when the delay constraint is relaxed.

5.4   Rainbow network flow problem

Rainbow network flow, which we first introduced in [28], is the problem of optimal routing of MDC packets in a general network. Unlike commodity flow, information packets can be duplicated at intermediate nodes. However, receiving duplicate descriptions is not beneficial in reconstructing the source. A node desires the *rainbow* effect by having as many distinct descriptions (or colors) as possible.

The RNF problem for balanced MDC is defined with the following inputs.

(1) $G\langle V, E\rangle$, a directed graph with a node set V and an edge set E.
(2) Two subsets $S = \{s_1, s_2, \ldots, s_{|S|}\}$, $T = \{t_1, t_2, t_3, \ldots, t_{|T|}\}$ of V representing the set of source and sink nodes respectively.
(3) A function $R : E \to \mathbb{R}^+$ representing the capacity of each link in G.
(4) A set $\chi \subset \mathbb{N}$ called the description set. We let $K \triangleq |\chi|$.
(5) An $r \in \mathbb{R}^+$ called the description rate.
(6) A non-increasing function $\delta : \{1, 2, \ldots, K\} \to \mathbb{R}^+$. The function δ is specified by the choice of balanced MDC. $\delta(k)$ is the reconstruction distortion when any subset of size k out of K possible descriptions is present at the decoder.
(7) A positive vector $\mathbf{p} = (p_t; t \in T)$ that weighs the importance of each sink node $t \in T$.
(8) A maximum tolerable delay, $n \in \mathbb{N}$.

This setup is the integral form of the problem and is called discrete Rainbow Network Flow (di RNF). The fractional form we call continuous Rainbow Network Flow (co RNF), and that is considered in Chapter 7.

We denote the above problem by $diRNF(G, R, r, K, n)$ where the dependence on $T, S, X, \mathbf{p}$ is understood. The goal of the RNF problem is to find routing paths from server nodes to sinks in which MDC description flow is a way to minimize weighted average distortion at the sink nodes.

To formulate the problem, we start by introducing the notion of flow of descriptions and in particular information duplication. A number of definitions are needed first. A flow path from $s \in S$ to $t \in T$ is a sequence of edges $w(s, t) = [(v_0 = s, v_1), (v_1, v_2), \ldots, (v_{m-1}, v_m = t)]$, such that $(v_i, v_{i+1}) \in E$ for $i = 0, 1, \ldots, m - 1$.

A rainbow network flow, denoted by $\dot{\alpha}(\dot{W}, f)$, consists of a set $\dot{W}$ of flow paths in G, and a so-called flow coloring function $f : \dot{W} \to \chi$ that assigns a description (or color) in χ to each flow path. An overhead dot (e.g., $\dot{O}$) denotes that the quantity of interest is related to a discrete version of the problem, as opposed to continuous Rainbow Network Flow considered in Chapter 7. This notation is respected in all other chapters as well.

For the flows in Fig. 5.2 for instance, $\dot{W} = \{[(1, 2)], [(1, 3)], [(1, 2), (2, 4)], [(1, 2), (2, 5)], [(1, 3), (3, 4)], [(1, 3), (3, 5)]\}$, and the flow coloring function f assigns:

$$f([(1, 2)]) = f([(1, 2), (2, 4)]) = f([(1, 2), (2, 5)]) = 1$$

and

$$f([(1, 3)]) = f([(1, 3), (3, 4)]) = f([(1, 3), (3, 5)]) = 2.$$

Let $\dot{\Phi}_E(e, \dot{W})$ and $\dot{\Phi}_V(v, \dot{W})$ be the sets of all colored flow paths in $\dot{W}$ that contain the link e or the node v, respectively.

For example, $\dot{\Phi}_E(e = (1, 2), \dot{W}) = \{[(1, 2)], [(1, 2), (2, 3)], [(1, 2), (2, 5)]\}$.

The spectrum of an edge $e \in E$, with respect to RNF $\dot{\alpha}$, is defined as:

$$\dot{\Psi}_E(\dot{\alpha}, e) \triangleq \bigcup_{w \in \dot{\Phi}_E(e, \dot{W})} f(w).$$

Likewise, the spectrum of a node v is defined as:

$$\dot{\Psi}_V(\dot{\alpha}, v) \triangleq \bigcup_{w \in \dot{\Phi}_V(v, \dot{W})} f(w).$$

The spectrum of a node indicates the set of descriptions available to the node. Likewise, the spectrum of an edge is the set of descriptions that flows into that edge.

In Fig. 5.2, for instance,

$$\dot{\Psi}_E((1, 2)) = \dot{\Psi}_E((2, 4)) = \dot{\Psi}_E((2, 5)) = \{1\}$$

and

$$\dot{\Psi}_E((1, 3)) = \dot{\Psi}_E((3, 4)) = \dot{\Psi}_E((3, 5)) = \{2\}.$$

The spectrum of the nodes $4, 5$ consists of both descriptions:

$$\dot{\Psi}_V(4) = \dot{\Psi}_V(5) = \{1, 2\}$$

and

$$\dot{\Psi}_V(2) = \{1\}, \dot{\Psi}_V(3) = \{2\}.$$

A rainbow network flow $\dot{\alpha}(\dot{W}, f)$ is said to be admissible with capacity function R, if and only if:

$$n^{-1} r |\dot{\Psi}_E(\dot{\alpha}, e)| < R(e) \qquad \forall e \in E. \tag{5.5}$$

The significance of this inequality is that it allows the duplication of a description by relay nodes. Therefore, two flow paths of the same color can pass through a link e, and yet consume a bandwidth of only r.

We denote the set of all admissible rainbow network flows by $\dot{\mathcal{F}}(G, R)$.

The rainbow flow plotted in Fig. 5.2 is admissible because at most one description with rate C is communicated over links $(2, 4), (2, 5), (3, 4), (3, 5)$. Note that the descriptions flowing into nodes 2 and 3 are duplicated.

For a given $\dot{\alpha}(\dot{W}, f)$, $|\dot{\Psi}_V(\dot{\alpha}, v)|$ is the number of distinct descriptions (out of a total of K such descriptions) available to node v. If the node t is a sink node, the reconstruction distortion at t will therefore be:

$$d_t = \delta(|\dot{\Psi}_V(\dot{\alpha}, t)|). \tag{5.6}$$

The weighted average distortion at all the sink nodes is then:

$$\overline{d}(\dot{\alpha}, \delta) = |T|^{-1} \sum_{t \in T} p_t \delta(|\dot{\Psi}_V(\dot{\alpha}, t)|). \tag{5.7}$$

The RNF problem is that of finding an admissible rainbow network flow that minimizes (5.7) for a given distortion metric δ.

The above problem is efficiently solvable for some classes of network topologies and distortion functions δ. To our special interest is the function $\delta^{(CRNF)}(k) = 1 - k/K$. For this choice of δ, the optimization reduces to maximizing the total number of distinct descriptions available to all sink nodes. As such, we call this particular RNF problem Cardinality RNF (CRNF).

Unfortunately, as will be proved in the next chapter, the CRNF problem is NP-Hard even on Directed Acyclic Graphs (DAGs). It does, however, have a binary integer program on DAGs. For tree topologies, the problem has a dynamic programming solution for arbitrary δ, as shown in the next chapter. We devote Chapter 6 to solving the RNF problem and presenting different complexity results.

5.5   Code design

The RNF problem is formulated assuming a given balanced MDC with distortion metric δ. Once δ is fixed, the RNF problem can, in principle, be solved. To find an optimized solution to the NASCC problem, we need to jointly optimize the choice of MDC (and hence the function δ) and RNF. This joint optimization, however, is computationally intractable. In the quest for a practical solution, we will separate the optimization into two parts: optimizing the flow using a generic distortion measure δ, and optimizing MDC design for the resulting flow.

The next step is to find a certain family of balanced multiple-description codes that are completely parameterizable, over which we can perform the code optimization. We choose a popular technique called Priority Encoding Transmission (PET) for MDC [24]. The following section introduces the class of MDC codes generated through PET and their parameterization. Finding an optimized flow (i.e., solving the RNF problem) is the subject of Chapter 6.

5.5.1   MDC using PET

The PET technique can produce any number of balanced multiple descriptions out of a *progressively encoded source stream* as illustrated in this subsection.

To make K descriptions of rate r bits per symbol, we proceed as follows.

Let $y = (y_i, i = 1, 2, \ldots, K)$ be a real-valued positive vector of length K such that $\sum_{i=1}^{K} y_i = 1$. Choose an integer m large enough so such that mr and my_l can be approximated by integers for all $l = 1, 2, \ldots, K$. Let $Y = (B_{ij}, i = 1, 2, \ldots, K, j = 1, 2, \ldots, m \cdot r)$ be a $K \times (m \cdot r)$ binary matrix.

Encode m samples of X into a progressive bitstream $(b_0, b_1, \ldots, b_{mrK})$.

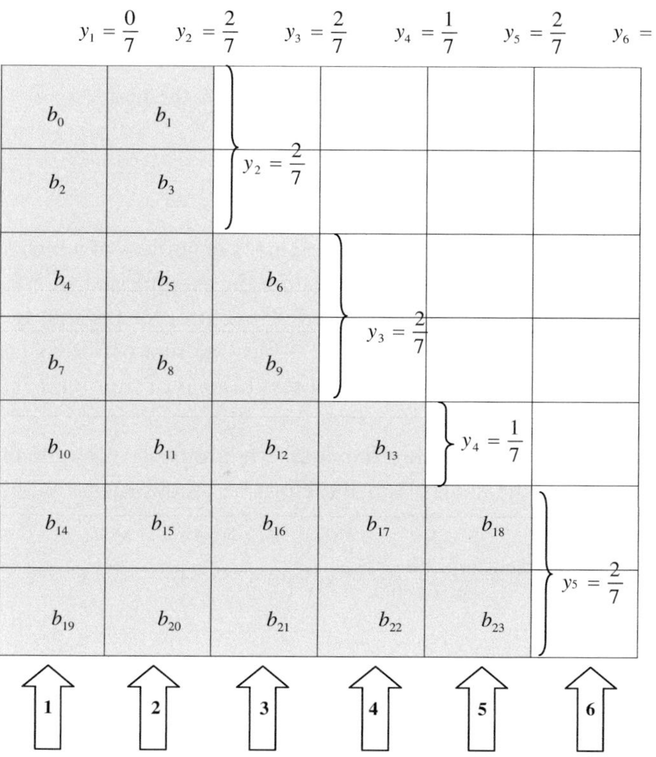

Figure 5.5 MDC produced with PET technique: The shaded cells corresponds to source bits from b_0 to b_{23}. The white cells correspond to parity bits of an erasure correction code corresponding to the source bits in the same row (e.g., a shortened Reed Solomon code or a Low Density Parity Check code). In this example, six different descriptions are produced (one for each column of the matrix). The MDC design is uniquely identified with a vector $\mathbf{y} = (0, 2, 2, 1, 2, 0)/7$. In the limit of large number of bits, the elements of y can be arbitrarily close to any real number

For $i = 1, 2, \ldots, K$ do the following:

For $l = 1, 2, \ldots, K$, let B_l, B_l' be sub-matrices of B consisting of:

$$B_l = \left[B_{ij};\ i = 1:l, \quad j = \sum_{k=1}^{l-1} m \cdot r y_k \ to\ \sum_{k=1}^{l} m \cdot r y_k \right]$$

$$B_l' = \left[B_{ij}';\ i = l+1:K, \quad j = \sum_{k=1}^{l-1} m \cdot r y_k \ to\ \sum_{k=1}^{l} m \cdot r y_k \right].$$

Therefore, matrix B_i contains $i \cdot m \cdot r \cdot y_i$ bits while B_i' has $(K - i) \cdot m \cdot r \cdot y_i$ bits. For $i = 1, 2, \ldots, K$, put the $i \cdot m \cdot r \cdot y_i$ bits of the progressive source code stream, from $b_{g(i)}$ to $b_{g(i)+i \times m \cdot r \cdot y_i}$ in B_i, where $g(i) = \sum_{k=1}^{i} k \cdot m \cdot r y_k$. In B_i', on the other hand, put parity symbols of a $(i \cdot m \cdot r \cdot y_i,\ K \cdot m \cdot r \cdot y_i)$ ideal *erasure correction* code corresponding to the bits in B_i.

This process is depicted in Fig. 5.5.

Now the descriptions consist of the K columns of the matrix B, each of $m \cdot r$ bits. The total source bits used equal $m \cdot r \sum_{k=1}^{K} k y_k$.

It is easily verified that given any $l \leq K$ descriptions, the first

$$\xi_l = \sum_{k=1}^{l} k \cdot m \cdot r \cdot y_k$$

bits of the source bit stream can be recovered. This is regardless of which l of the total of K descriptions are available. To see why, consider the example in Fig. 5.5. Let's assume that two columns of the matrix are recovered. Therefore, for the first two rows of the matrix, two symbols are correctly recovered. Thus, the first two rows can be decoded to obtain the source bits in the rows one and two (a total of four bits). This is equal to $r(y_1 \times 1 + 2 \times y_2) = 7 \cdot 4/7 = 4$.

For large enough m and assuming the source is progressively refinable, given any k distinct descriptions, the source can therefore be reconstructed within an average distortion:

$$D_X(\xi_k/m) = D_X\left(r \sum_{l=1}^{k} l y_l\right). \tag{5.8}$$

It can be shown [29] that any real valued i.i.d. source is almost progressively refinable. More precisely, for any real valued source with rate-distortion function $D(z)$, a progressively refinable source code exists with a rate-distortion performance $D_{prog}(z) \leq D(z - 1/2)$ for all $z > 1/2$.

Excellent progressively refinable source coding schemes exist for most multimedia signals (and in particular images [26, 30]). For the rest of this chapter we assume that the source has been progressively encoded with a rate-distortion function $D_X(\cdot)$.

5.5.2   Optimizing code for a fixed rainbow flow

Vectors $\mathbf{y}$ parameterize the space of all MDCs that can be generated through PET. It is over this space that we will carry out our code optimization.

For any admissible flow $\dot{\alpha}$, define the rainbow flow vector (RFV), $\dot{\mathbf{q}}(\dot{\alpha}) = (\dot{q}_t; t \in T)$ such that:

$$\dot{q}_t = n^{-1} r |\dot{\Psi}_V(\dot{\alpha}, t)|. \tag{5.9}$$

In other words, $\dot{q}_t$ is the volume of distinct descriptions available to a sink node t per unit time. For future use, let $\dot{\mathcal{Q}}(G, R, r, K, n) \subset \mathbb{R}^{|T|}$ denote the set of all RFVs $\dot{\mathbf{q}}(\dot{\alpha})$ that result from all admissible flows $\dot{\alpha} \in \dot{\mathcal{F}}(G, R)$.

For a given flow with RFV of $\dot{\mathbf{q}} = (\dot{q}_t; t \in T)$, the weighted average reconstruction distortion at all sink nodes can be written as:

$$\bar{d}(\mathbf{y}, \dot{\mathbf{q}}) = |T|^{-1} \sum_{t \in T} p_t D_X\left(n^{-1} r \sum_{l=1}^{n\dot{q}_t/r} l y_l\right). \tag{5.10}$$

Note that $n\dot{q}_t/r$ is an integer from (5.9).

An optimized MDC can be found by solving the following problem:

$$\min_{\mathbf{y}\succ 0,\|\mathbf{y}\|_1=1} \sum_{t\in T} p_t D_X \left(r \sum_{l=1}^{n\dot{q}_t/r} ly_l \right). \tag{5.11}$$

The rate-distortion function $D_X(\cdot)$ is a convex function under ideal conditions [31] and is very close to being convex for most state-of-the-art multimedia source codes [30], [32]. When $D_X(\cdot)$ is a convex function, the objective in (5.11) is the sum of convex functions of linear combination of variables. Therefore, (5.11) is a convex optimization problem with linear constraints and can be solved efficiently using standard tools [33].

5.5.3 Discrete optimization approaches

The optimization approach in (5.11) relied on a number of assumptions. First, we have assumed that the source is progressively encodable with fine granularity. In other words, the truncation of the code stream is possible at any point. This is in fact true for the state of the art image coders such as JPEG2000 [26] and SPHIT [30]. Second, the vector $\mathbf{y}$ can assume any real number in the range $[0, 1]$. In practice, however, one should consider the discrete nature of information in design, the smallest units being bits, and usually bytes. This problem, at least in theory, is alleviated by encoding large enough samples of the source ($m \gg 1$). Third, we have assumed that the rate-distortion function is convex. With these assumptions, the optimization problem in (5.11) can be conveniently solved.

It turns out that the code design problem through PET can still be efficiently solved, through dynamic programming, even if neither of these assumptions holds. In fact, with a change of variables, we can turn our design problem into an equivalent formulation for which a large body of literature exists.

PET has been used in multimedia applications in which data packets are likely to be lost. To model packet losses, it is often assumed that a given user receives a total of k descriptions with probability $u(k)$. MDC is then optimized to minimize the expected distortion over all receivers.

Previous works in the literature consider a different, but equivalent, parameterization of all designs achievable through PET. Let π_i denote the number of source bits in the row i of the submatrix B_i. Given any subset of size k out of the total of K descriptions, one can recover:

$$\theta_k = m^{-1} \sum_{i:\pi_i \leq r \cdot m \cdot k} \pi_k.$$

Therefore, the average reconstruction distortion at a given user is:

$$d(\pi_1, \pi_2, \ldots \pi_M) = \sum_{k=1}^{K} z(k) D_X(\theta_k),$$

where M is the number of layers (rows of the matrix). Given $M, K, u(.), D_X(.)$, the goal of the optimization problem is to minimize $d(\pi_1, \pi_2, \ldots \pi_M)$ given the constraint that $\pi_1 \leq \pi_2 \ldots \leq \pi_M$.

> **Box 5.1** Proposed algorithm for finding an optimized solution to NASCC problem
>
> **Optimized Solution to NASCC Problem**
>
> (1) Choose a value of r and the total number of descriptions K.
> (2) Solve the CRNF problem by methods in Chapter 6 to obtain the optimal flow $\dot{\alpha}^{(CRNF)}$.
> (3) Find the RFV, $\dot{\mathbf{q}}^{(CRNF)}$.
> (4) Solve the convex optimization problem of minimizing:
>
> $$\overline{d}(\mathbf{y}, \dot{\mathbf{q}}^{(CRNF)}) = |T|^{-1} \sum_{t \in T} p_t D_X \left(r \sum_{l=1}^{n\dot{q}_t^{(CRNF)}/r} ly_l \right)$$
>
> under the constraint $y_l \geq 0, \sum_{l=1,2,\dots,K} y_l = 1$.
> (5) Construct the MDC streams for the vector
> $\mathbf{y} = (y_1, y_2, \dots, y_K)$ with the method in Section 5.5.1.

In our problem, $u(k)$ is equivalent to the fraction of sinks that receive exactly k distinct descriptions.

Many algorithms have been proposed for solving the problem, the most successful of which is due to Dumitrescu, Wu, and Wang [34], which solved the problem for integer π_i in time $O(K^2 M^2)$ regardless of the convexity of $D_X(\cdot)$ and in time $O(KM^2)$ if $D_X(\cdot)$ is convex. Other works that either require the convexity assumption or have to relax the integer assumption on π_i's include [35–39]. We provide a rather complete account of algorithms for MDC design in Chapter 8.

We now have a systematic way to find an optimized solution to the NASCC problem. We start by solving the CRNF problem (i.e., the version of RNF problem for $\delta^{(CRNF)}(k) = 1 - k/K$. We then optimize the MDC with methods in Section 5.5.2. Our algorithm is summarized in Box 5.1.

5.6   Numerical simulations

The proposed JSNC approach is tested on a number of randomly generated DAGs. For DAGs, CRNF admits a linear binary program. The simulation results examined in this section are based on solving this problem on relatively small network sizes using a numerical binary optimization package. We first solve the instance of the CRNF problem and then optimize the MDC streams for the resulting RFV. Theoretical and practical considerations for solving CRNF on DAGs and other graphs are given in Chapter 6.

5.6.1   Network simulation setup

We produce a family of DAGs motivated by a simplistic model for the growth of peer-to-peer networks. We start from a single node. At each step, a new node is added. For some prespecified value of P, an integer $1 \leq p \leq P$ is randomly chosen. Then, p nodes are chosen at random with replacement from the existing nodes, and a link is made from each of the p chosen nodes to the new node. Once the network grows to N nodes, we assign a capacity $C(e)$ to each link e. $C(e)$ is a random integer between $C_{min} = 1$ and C_{max} for some choice of maximum capacity C_{max}. We take the rate of the descriptions to be one. Note that all nodes have an in-degree of at most P. While the average out-degree converges to $(M + 1)/2$, for large N, its distribution can be shown to have an exponentially decaying tail [40].

The results in this section are obtained by solving the 0-1 linear integer program discussed in Chapter 6 using a commercial package for solving integer programs [41]. For small values of K (up to $K = 8$), our package performed extremely well for network sizes of up to 2000 nodes. When K and the maximum capacity C_{max} increase, however, we observed an exponential increase in the computation time. In those cases, the algorithm in Chapter 6 can be used to find provably good approximate solutions to the CRNF problem on DAGs.

5.6.2   Effect of the number of descriptions

It can be shown that for a fixed source and network, increasing the number of descriptions K will not increase the overall distortion (see Chapter 7). To keep the running time manageable when using our numerical optimization package, we always start with the smallest possible K, and increase K until the overall distortion does not decrease any more.

For a network of size $N = 50$, with nodes of maximum in-degree $P = 3$ and $C_{max} = 3$, the overall distortion $\bar{d}$ is optimized by first solving the instance of CRNF problem and then optimizing the MDC. In all cases, description rates are $r = 1$. The optimization is repeated for an increasing number of descriptions K. The distortion converges to its final value for $K = 6$ as shown in Fig. 5.6. The overall optimization took less than 5 seconds.

5.6.3   The effect of network size

To see the effect of the network size, we have provided simulation results for fixed $K = 4$, $C_{max} = 3$, and $P = 3$ and different network sizes. For the above family of random networks, the increase in the network size will offer greater path diversity to the more recently created nodes, which are at the bottom of the network hierarchy. Therefore, as the network size grows, the fraction of nodes that receive a higher number of descriptions increases (see Fig. 5.7), which leads to a decrease in the overall average distortion. The average signal to noise ratio (SNR) resulting from our design is reported in Fig. 5.8. For comparison, the SNR obtained from separate source and network coding is also depicted. Our approach improves the average SNR by up to 3 dB.

Table 5.1 shows the optimal design parameters for the PET code for one particular realization of the network sizes $N = 50, 100, 200$. As the network size grows, a larger

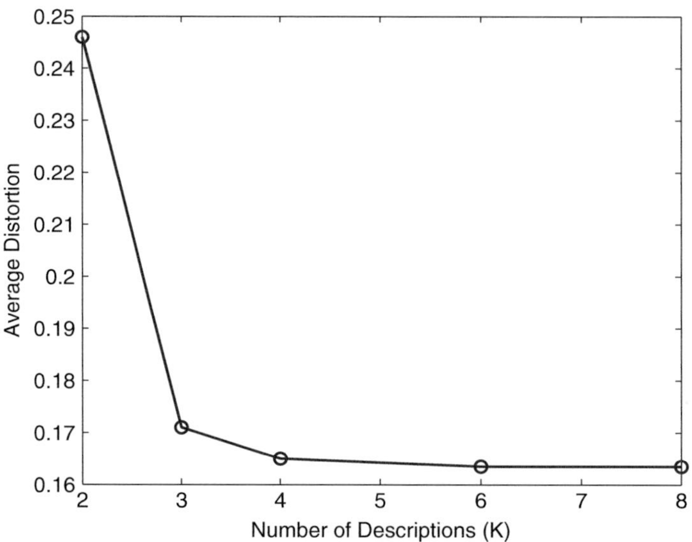

Figure 5.6 Distortion as a function of the number of descriptions: description rate is $r = 1$, the network has $N = 50$ nodes, and $C_{max} = 3$ and $P = 3$

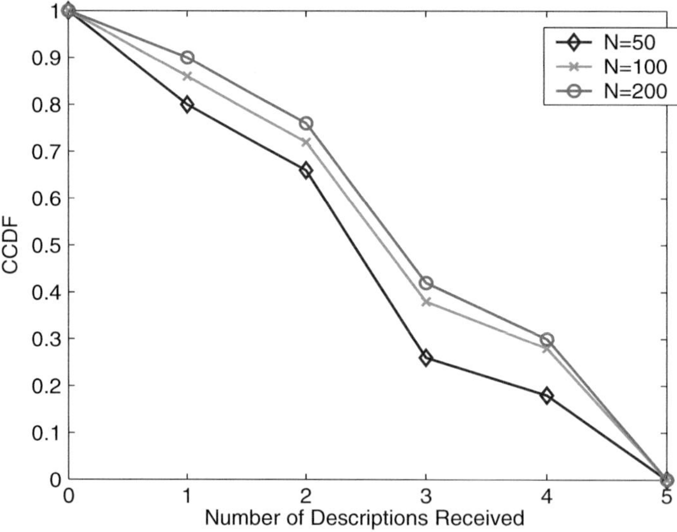

Figure 5.7 The fraction of nodes that receive at least k distinct descriptions as a function of k for different network sizes (i.e., complementary cumulative distribution function or CCDF): We have $C_{max} = 3$, $P = 3$, and $K = 4$. Therefore, $k = 0, 1, \ldots, 4$. For larger network sizes, due to an increase in the number of available paths, the fraction of nodes that receive a higher number of distinct descriptions increases

fraction of nodes will have access to a higher number of descriptions. This is reflected in the optimal solution of y_l by noting that the optimal $\mathbf{y}$ has more significant non-zero values for larger l when the network size grows.

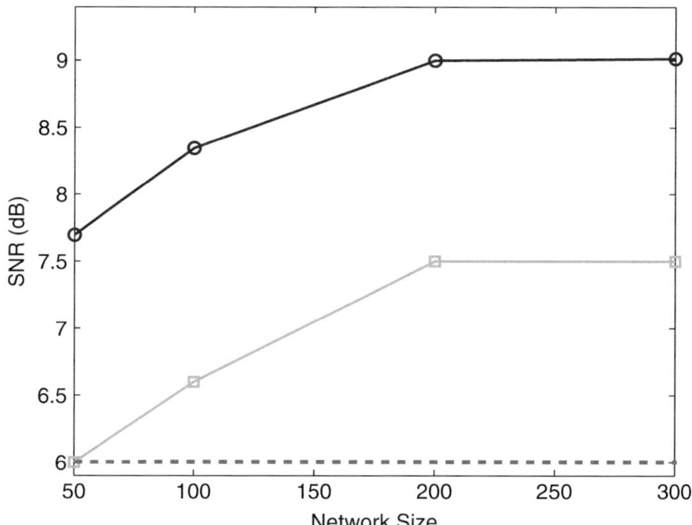

Figure 5.8 The average distortion as a function of the network size: For each network size, the results are an average of 3 trials. The circles correspond to the case where the path optimization is done with optimization methods in Chapter 6. Squares correspond to the case where path optimization is done through a GCB algorithm in Section 5.6.4. The dashed line is the average SNR achievable with separate source and network coding

Table 5.1 The optimal vector **y** for different network sizes. $C_{max} = 3$, $K = 4$, and $P = 3$

N	$\overline{d}^*$	y_1	y_2	y_3	y_4
50	0.1604	0.0	0.7069	0.2067	0.083
100	0.143	0.0	0.445	0.396	0.0160
200	0.126	0.0	0.204	0.556	0.114

5.6.4   The effect of the performance of path optimization algorithms

The quality of the solutions obtained by the proposed method greatly depends on our ability to find good diversity routes. The algorithms proposed in Chapter 6, however, are mostly global optimization algorithms. In many applications, global optimization strategies might not be feasible. One might therefore need to consider local optimization algorithms. For the sake of comparison, we considered Greedy Color Broadcast (GCB), a local heuristic optimization algorithm.

The GCB works as follows. Initially, all descriptions are owned by server nodes only. Let v' be a neighbor of v, i.e., $(v, v') \in E$. Let $\mathcal{K}(v, v')$ be the set of all descriptions that are owned by v but not v'. At each step, node v sends a description randomly chosen from $\mathcal{K}(v, v')$ to v'. This process is repeated until no more descriptions can be communicated.

We found the performance of our design approach when routes are optimized through GCB. While the performance is still superior to separate source and network coding, a careful, global optimization of routes provides a much better solution, as illustrated in Fig. 5.8.

Finally, we should note that the interaction of the proposed design approach with the underlying network topology is a rich and fascinating one. For instance, a plausible question is, what is the limit of average distortion for a family of random networks as the network size goes to infinity? What is the minimum number of descriptions after which the average distortion does not decrease further? These and many other questions perhaps deserve a separate treatment elsewhere.

5.7   Concluding remarks

This chapter proposed a concrete algorithmic approach to lossy source communication in general networks. The main idea was to exploit diverse communication paths from a server node to different sink nodes. This is made possible through multiple-description coding. The challenging problems are (1) effectively optimizing the flow of the descriptions, and (2) at the same time, designing the descriptions to minimize the reconstruction distortion for the corresponding flows. Since the joint optimization is intractable, we devised a two-step approach which consists of optimizing the flows to find one with *good* diversity, and then optimizing the codes while keeping the flows fixed.

While designing the codes for a fixed "good" flow is a convex optimization problem, finding a good flow is not necessarily straightforward. In the following chapter we will consider aspects of this optimization problem. In particular, we will find a $0 - 1$ linear integer program representation of the CRNF problem on DAGs, which can be solved for moderately large networks using existing numerical optimization packages. The actual problem itself is proved to be NP-Hard.

In our designs, the description rates r and the tolerable delay n were given as inputs. In general, the designer might have control over these parameters. The effects of these parameters on the quality of the design are investigated in Chapter 7. We devote the next chapter to different algorithms for solving the RNF problem and their complexity issues.

6   Solving the rainbow network flow problem

The methods proposed in the previous chapter rely on our ability to find a, perhaps approximate, solution to the RNF problem, and in particular the CRNF problem. Once such a solution is found, the design of the MDC codes with PET technique is straight-forward. This chapter is a more detailed algorithmic account of RNF and in particular the CRNF problem.

We start by proving the CRNF problem to be NP-Hard in Section 6.1, even on Directed Acyclic Graph (DAG) network topologies. In Section 6.2, we show that the CRNF problem can be posed as an integer linear program for DAGs. The problem can therefore be solved for moderate network sizes using existing numerical optimization packages.

For large networks, the results in Section 6.1 suggest that the exact solution to the linear integer program formulated in Section 6.2 can not be found efficiently. While the CRNF problem is NP-Hard even on DAGs, we find a polynomial-time solution for a generalized tree topology. If the network graph can be appropriately decomposed into tree components, then a dynamic programming algorithm can be applied to solve the CRNF problem exactly for a general distortion function $\delta(k)$. The development of this algorithm is the subject of Section 6.3.

6.1   Complexity results of the CRNF problem

In this section, we prove that the CRNF problem is NP-Hard. The proof is constructed by reducing the well-known NP-hard problem of graph 3-colorability [42] to a special instance of CRNF on a DAG, where there is only a single server node.

Given a 3-colorable graph $G = \langle V, E \rangle$, we create an instance of CRNF, which is a directed acyclic graph $G' = \langle V', E' \rangle$. The reduction is carried out as follows.

(1) Initialize $V' = \phi$ and $E' = \phi$;
(2) add s^* to V' and let $S = \{s^*\}$;
(3) for each $v \in V$, add a v' to V' and (s^*, v') to E';
(4) for each edge $(u, v) \in E$, add a vertex v_{uv} to V' and add (u', v_{uv}) and (v', v_{uv}) to E' (vertex v_{uv} corresponds to an edge in G);
(5) for each member of the set $\{u, v, w \in V | (u, v), (v, w), (w, u) \in E\}$, add v_{uvw} to V', add (u', v_{uvw}), (v', v_{uvw}), (w', v_{uvw}) to E', and v_{uvw} to T (vertex v_{uvw} corresponds to a triangle in G);

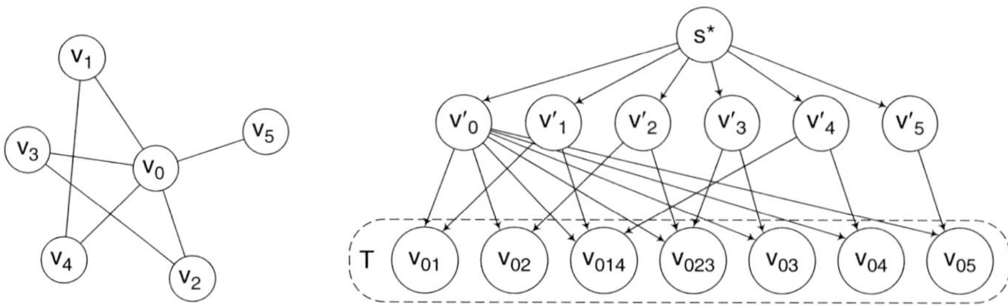

Figure 6.1 The reduction from an instance of graph 3-colorable: (a) to an instance of CRNF (b) all edges have capacity 1

(6) for each $v' \in V'$ add (s^*, v') to E';
(7) set capacities of all edges to 1;
(8) set all description rates to be 1;
(9) let the MDC code have 3 descriptions.

The above reduction can be effected in polynomial time. An example of this reduction is illustrated in Fig. 6.1. As is evident in Fig. 6.1, the reduction produces an instance of the CRNF problem on a DAG topology.

To finish the proof, we need the following lemma.

LEMMA 6.1   *The graph $G\langle V, E \rangle$ is 3-COLORABLE if and only if the transformed CRNF problem has a solution equal to the sum of the number of incoming edges of T.*

Proof $\Rightarrow$: Assume the graph $G = \langle V, E \rangle$ is 3-colorable by function $C : V \rightarrow \{1, 2, 3\}$. For each $v \in V$, let $v' \in V'$ have the description i from s^*, if $C(v) = i$, through the only edge connecting v' and s^*. For each $v' \in V'$, the only description is broadcast along all outgoing edges.

Each vertex in T corresponds either to an edge or to a triangle in G. If a vertex $t \in T$ corresponds to an edge in G, the two incoming edges must deliver to t different descriptions, otherwise the assumption is contradicted. If a vertex $t \in T$ corresponds to a triangle in G, the three incoming edges must deliver to t three distinct descriptions, or again, a contradiction of the assumption results. Since for each client, all incoming edges deliver distinct descriptions, we have the optimal solution to the CRNF problem.

$\Leftarrow$: Assume that in the CRNF problem, the total number of distinct descriptions communicated to the sink nodes is equal to the sum of the number of incoming edges of T. For this to hold, for each client, all incoming edges must deliver distinct descriptions. Since each server has only one description delivered from s^*, we use that description to color the corresponding vertex of G. Thus the graph G is 3-colorable. $\square$

The decision version of the CRNF problem is simply to determine whether the optimal value is equal to an arbitrary integer K. Clearly, the corresponding decision problem is in NP. And from Lemma 6.1, Theorem 6.1 follows.

THEOREM 6.1   *CRNF is NP-Hard even when there is a single-server node and the underlying topology is a DAG.*

6.2 A binary integer program for CRNF on directed acyclic graphs

The results in the previous section state that the CRNF problem is NP-Hard even for DAG topologies. Nevertheless, for DAGs, the CRNF problem can be posed as a 0-1 linear integer program. This allows one to find the optimal solution for moderate network sizes using numerical packages.

6.2.1 Formulation

Let the binary variable $x_{v,v'}^k$ be 1 if description k flows into the edge (v, v') and 0 otherwise. Likewise, for each node v, the binary variable $y_v^k \in \{0, 1\}$ indicates whether description k is received by v. For $s \in S$ we have $y_s^k = 1$ automatically.

Let $O(j)$ for $j \in V$ be the set of all parent nodes of a node j. Since each node $j \notin S$ has to receive the descriptions from a parent node, we have:

$$y_j^k \leq \max_{i \in O(j)} x_{i,j}^k.$$

This is a crucial equation. It indicates that the rate of useful information about a description k received by a node j from all its parent nodes is equal to the maximum of the rate of information available to the parent nodes. In other words, the information rates about a single description k do not add up (unlike a physical flow). But, any network node can multicast a description by duplicating it (again, unlike a physical flow that cannot be duplicated). However, node i has to first possess a copy of description k before transmitting duplicated copies to different destinations, if required. Thus

$$\forall j \ \ x_{i,j}^k \leq y_i^k. \tag{6.1}$$

The edge capacity constraint is:

$$\sum_k x_{i,j}^k \leq \frac{R(i,j)}{r}. \tag{6.2}$$

For each sink $t \in T$, the number of distinct descriptions received is $\sum_k y_t^k$. Therefore, the total fidelity is

$$L \triangleq \sum_{t \in T} \sum_k y_t^k.$$

Then the CRNF problem can be cast into the following linear integer programming problem:

$$\max \quad L \triangleq \sum_{t \in T} \sum_k y_t^k \tag{6.3}$$

$$\text{subject to} \quad x_{i,j}^k \leq 1 \tag{6.4}$$

$$y_j^k \leq \max_{i \in O(j)} x_{i,j}^k \quad \text{if } j \notin S$$

$$y_s^k = 1 \quad \text{if } s \in S$$

$$x_{i,j}^k \leq y_i^k$$

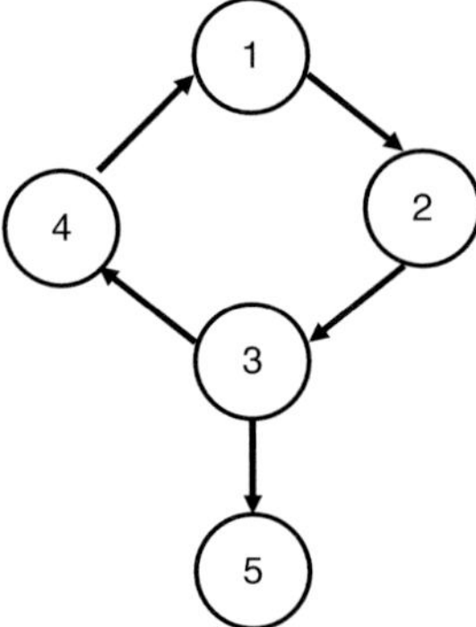

Figure 6.2 When duplication is allowed, the loops can generate unreal flows while all the flow constraints are satisfied

$$\sum_{k} x_{i,j}^{k} \leq \frac{R(i,j)}{r}$$

$$y_{j}^{k} \in \{0, 1\}, \quad x_{i,j}^{k} \in \{0, 1\}.$$

6.2.2　The DAG requirement

The above formulation works only for DAGs. The problem is that the ability to duplicate data might lead to a situation in which information is generated by a node that is not a server node. One example of this is depicted in Fig. 6.2. In this case, a description circles the loop $1 \to 2 \to 3 \to 4$, while "leaking out" a copy to node 5. Note that all the flow constraints in (6.3) are still satisfied. If node 5 is a sink, it will receive a copy of the description without requiring any of the nodes 1 to 4 to be a server node. This scenario can obviously be avoided if the underlying topology is a DAG.

6.3　Solving CRNF on tree-decomposable graphs

Before presenting our algorithm, a number of points are worth mentioning. First, while this thesis has primarily considered directed graphs, the algorithm developed in this section works for both directed and undirected networks with tree topologies. As such, we state the algorithm assuming undirected links. The same algorithm can be applied to the case of directed edges with virtually no modifications. Second, the algorithm can solve the CRNF problem for *any* balanced MDC distortion function $\delta(k)$, including the one corresponding to CRNF ($\delta^{(CRNF)}(k) = 1 - k/K$).

Tree-decomposable topologies: The algorithm developed in this section works for a class of so-called tree-decomposable topologies, including tree topologies as its special case. The notion of tree decomposability is illustrated in Fig. 6.3. We break the network at each server node $s \in S$. A server node with m connections can be decomposed into m disjoint server nodes, each connected to one of the m links. Figure 6.3 shows the

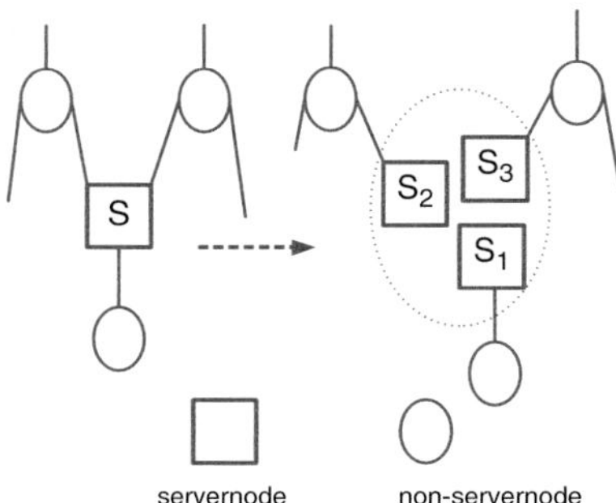

Figure 6.3 Tree-decomposability: Each server node with degree m can be decomposed into m independent server nodes. The figure shows the procedure for a server node s with degree $m = 3$

decomposition for a server node s with $m = 3$ connections. As a result of splitting the server nodes, the network is fragmented into a number of connected components. Since each server node has all the descriptions, it does not need to forward any description from any other nodes in the optimal solution. Therefore, the optimization for CRNF can be carried out for each of the connected components separately.

We call a network tree-decomposable if all the connected components resulting from decomposing the server nodes are trees. Trivially, every tree topology is also tree decomposable, but not vice versa. In the remainder of this section, we develop an algorithm that solves the RNF problem on tree topologies. This algorithm, therefore, can be used to efficiently solve the RNF problem on any tree-decomposable topology. Therefore, in the rest of this section, we assume that the graph $G\langle V, E\rangle$ is a tree.

Before getting to the description of the algorithm, let's clarify what we mean by a tree topology when the graph is directed. Let $G = \langle V, E\rangle$ be a directed graph. We convert it to an undirected graph $G' = \langle V, E'\rangle$ as follows: $(u, v) \in E'$ if and only if $(u, v) \in E$ or $(v, u) \in E$. If the new graph G' is a tree, we say that G has a tree topology.

We develop our algorithm for binary trees first, and then extend it to arbitrary trees. We can make G a binary tree rooted at an arbitrary leaf node $v \in V$ by inserting a dummy vertex v^* and a dummy edge (v, v^*) into G. Note that with insertion of dummy leaf nodes we can convert any rooted binary tree to one in which all internal nodes have exactly two children (one of which might be a dummy node).

Our proposed algorithm has two separate parts. The first, and the main, part is a dynamic program that calculates the optimal number of distinct descriptions that should flow into any given link in either direction. It leaves the actual assignment of descriptions to flow paths unspecified. Then, through a straightforward algorithm called flow coloring, the descriptions that should flow in each link are specified.

6.3.1　Calculation of the optimal flows

For each internal node u, let u' be the parent node of u. For the edge $j = (u', u)$, let $f_+ \in \mathbb{I}$ denote the number of distinct descriptions that flow from u' to u. Likewise, let $f_- \in \mathbb{I}$ denote the number of distinct descriptions that flow from node u to its parent node u'.

Our goal is to propose an algorithm to calculate, for every link, the optimal number of descriptions that should flow in either direction (e.g., f_+, f_- for the link (u', u)).

We take a dynamic programming (DP) approach to solving the problem. Let $DP(u, f_+, f_-)$ be the minimum sum of the distortions at the sinks in the subtree rooted at u, given that the number of distinct descriptions received by u from its parent is f_+. Let v, w be the two children of u. Let f'_+ denote the number of distinct descriptions that flow from node u to its left child v. Similarly, let f'_- be the number of distinct descriptions that flow from node v into node u. f''_+, f''_- are defined for the right child w in the same way. The dynamic programming algorithm relies on a recurrence relation to compute $DP(u, f_+, f_-)$ from $DP(v, f'_+, f'_-)$ and $DP(w, f''_+, f''_-)$.

To derive the recurrence, we need to differentiate the following four cases, depending on the position and function of a network node u: whether u is a server node, a pure relay node, or a relay node which is also a sink.

Case 1. u is a leaf and a server node

In this case, the descriptions should only flow out of the server node, while respecting the capacity of the links. Also, the number of distinct descriptions is always bounded by K. Therefore,

$$DP(u, f_+, f_-) = \begin{cases} 0, & \text{if} \quad f_+ = 0 \ and \ f_- \leq \min\left\{\frac{R(u,u')}{r}, K\right\} \\ +\infty, & \text{otherwise} \end{cases} \tag{6.5}$$

Case 2. u is both a leaf node and a sink

In this case, no description is required to flow out of the node u. Therefore,

$$DP(u, f_+, f_-) = \begin{cases} \delta(f_+), & \text{if} \quad f_- = 0 \ and \ f_+ \leq \min\left\{\frac{R(u,u')}{r}, K\right\} \\ +\infty, & \text{otherwise} \end{cases} \tag{6.6}$$

Case 3. u is a pure relay node (which is necessarily an internal node)

In this case, the node u does not directly contribute to the overall distortion (because it is not a sink node). Therefore, the following recurrence relation should hold:

$$DP(u, f_+, f_-) = \min\left(DP\left(v, f'_+, f'_-\right) + DP\left(w, f''_+, f''_-\right)\right) \tag{6.7}$$

for all f'_+, f'_- and f''_+, f''_- satisfying the following conditions.

(1) The capacity constraints on all edges should be satisfied. Thus,

$$f_+ + f_- \leq \min\left\{\frac{R(u, u')}{r}, K\right\}$$

$$f'_+ + f'_- \leq \min\left\{\frac{R(u, v)}{r}, K\right\} \tag{6.8}$$

$$f''_+ + f''_- \leq \min\left\{\frac{R(u, w)}{r}, K\right\}.$$

(2) Since the node u is not a server node, the number of distinct descriptions that flow out of u cannot be more than the number of distinct descriptions it receives

from other nodes. Therefore, $f \leq \min\left\{f'_- + f''_-, K\right\}$, $f'_+ \leq \min\left\{f_+ + f''_-, K\right\}$ and $f''_+ \leq \min\left\{f_+ + f'_-, K\right\}$.

Case 4. u is an internal node and a sink

This case is similar to Case 3, except that the recurrence relation in (6.7) should be modified into:

$$DP(u, f_+, f_-) = \min \left(DP\left(v, f'_+, f'_-\right) + DP\left(w, f''_+, f''_-\right) \right.$$
$$\left. + \delta \left(\min\left\{f_+ + f'_- + f''_-, K\right\} \right) \right). \tag{6.9}$$

The constraints of Case 3 should be satisfied in Case 4 as well.

The optimal number of distinct descriptions that have to flow in each link in either direction can now be found by solving for $DP(v_o, 0, 0)$, with v_o being the root of tree network G. To do this, in post-order traversal of the tree nodes from the leaves to the root, the algorithm recursively solves the problem for the subtrees rooted at the traversed nodes. At each step, the current node u is classified into one of the above four cases. The corresponding recurrence relation is used to compute the optimal solution $DP(u, f_+, f_-)$ for the subtree G_u rooted at u from the optimal solutions $DP\left(v, f'_+, f'_-\right)$ and $DP\left(w, f''_+, f''_-\right)$ for the left and right subtrees of G_u.

Up until now, we have only considered the number of descriptions that should flow in a given link, and we have left the choice of those descriptions unspecified. For all $(u', u) \in E$, we let $f(u', u)$ indicate the number of distinct descriptions that should flow from node u to node u'. Given this information, the choice of which description to flow in which edge can be made through a recursive algorithm, called flow coloring algorithm, which is discussed next.

6.3.2   Flow coloring

The dynamic program in Section 6.3.1 only specifies, for every link, the number of distinct descriptions that should flow in either direction (i.e., $f(u, u')$).

Let R_S be the sum of the outgoing capacities of all server nodes, i.e.,

$$R_S = \sum_{s \in S} \sum_{(s,v) \in E} R(s, v). \tag{6.10}$$

If $K \geq \lfloor R_S/r \rfloor$, a simple, greedy method can be used to specify which descriptions should flow in each link, as is described next.

For a server node $s \in S$ and a description $k \in \{1, 2, \ldots, K\}$, a Color Broadcasting Round ($CBR(s, k)$) is defined as follows. If $f(s, u) > 0$, the server node s sends the description k to its neighbor u (note that due to the tree-decomposition, the server node has exactly one neighbor). The value of $f(s, u)$ is then decremented by one. The procedure continues recursively, i.e., when a node u receives a description k, it passes a copy to each of its neighbors $u' : (u, u') \in E$ for which $f(u, u') > 0$. The counter $f(u, u')$ is then decremented by one. This establishes a flow path from the server node s to any node that receives a copy of description k.

Let $\mathcal{K}$ be the set that contains the descriptions that should be assigned to flow paths. Initially $\mathcal{K} = \{1, 2, \ldots, K\}$. To specify all paths, for each server node s, we call the procedure CBR(s, k) for some random $k \in \mathcal{K}$. By doing so, we essentially specify a spanning tree, rooted at the server node s, in which description k should flow. When CBR(s, k) is called, the description k is removed form the set $\mathcal{K}$. This is continued until no other description can flow out of s. The same procedure is then repeated for all server nodes in S.

The total number of descriptions that can flow out of s is at most

$$\left\lfloor \frac{\sum_{u:(s,u)\in E} R(s, u)}{r} \right\rfloor.$$

Thus, the above algorithm works as long as $K \geq \lfloor R_S/r \rfloor$. As will be shown in Chapter 7, as far as the NASCC problem is concerned, there is no loss in performance if one increases the number of available descriptions. Therefore, at least theoretically, one can always choose a large enough value for K to distinguish all descriptions that flow into any given edge.

THEOREM 6.2   *When $K = \lceil R_S/r \rceil$, the proposed algorithm finds the optimal solution of RNF problem for networks of tree topology. Its time complexity is*

$$O\left(|V| \times \left[\left(\frac{R_{max}}{r} \right)^3 + |S|^2 \cdot \frac{R_{max}}{r} \right] \right)$$

where R_{max} is the maximum link capacity in the network.

Proof   The optimality of the algorithm follows immediately from the fact that all possible cases of the dynamic program are covered through Cases 1 to 4 in Section 6.3.1. We still need to show that the optimal value flows calculated through the dynamic program are actually feasible. More precisely, for each $(u, u') \in E$, our proposed dynamic program calculates $f(u, u')$, the optimal number of distinct descriptions that should flow from a node u to u'. We still need to show that a flow coloring procedure exists that can actually deliver $f(u, u')$ distinct descriptions from u to u'. This is ensured with the flow coloring algorithm in Section 6.3.2. Since each description k flows out of a single server, node u can receive k only from one of the subtrees connected to it (see Fig. 6.4). As such, node u will never receive a duplicate description.

The time complexity of the dynamic program is dominated by the cost of the minimization task of (6.9) under the constraints in (6.8). Since the maximum capacity is R_{max}, for each edge (u', u), there are at most R_{max}/r choices for f_+, f_- that satisfy the constraints in Case 3. Similarly, there are at most R_{max}/r choices for f'_+, f'_- and f''_+, f''_- as well. Therefore, for each node u, there are at most $(R_{max}/r)^3$ valid combinations of $f_+, f_-, f'_+, f'_-, f''_+, f''_-$. As such, the time complexity of calculating the optimal flows is $O\left(|V| \times \left(\frac{R_{max}}{r} \right)^3 \right)$. The flow coloring algorithm runs in time $O(|V| \times |S| \times K)$. The assertion on time complexity follows by noting that $R_S \leq |S| \times R_{max}$ and using the value of K in the statement of theorem. $\qquad\square$

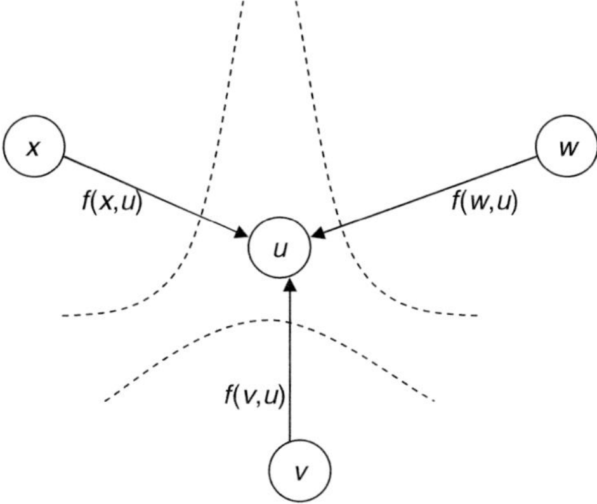

Figure 6.4 A node u receives distinct descriptions from each of its neighbors: Each description flows out of only one server. Thus, the node u can only receive a description k from the sole subtree to which the corresponding server node belongs

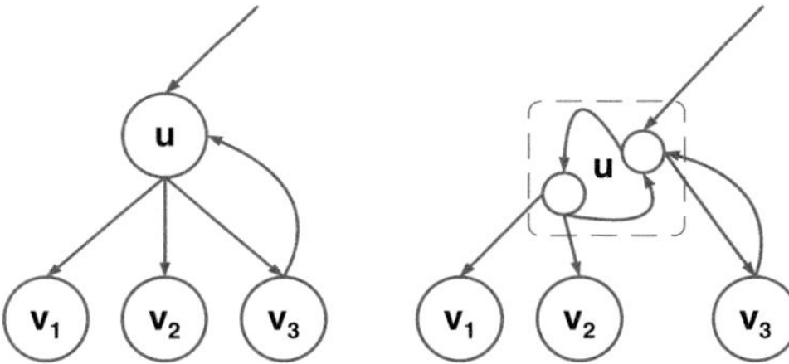

Figure 6.5 A general tree can be reduced to a binary tree

The algorithms developed so far considered binary trees only. For any tree with degree greater than 3, we can convert it to a binary tree. Each degree d node is split into $d - 2$ nodes, and the edges connecting these nodes have unlimited capacity. An example is given in Fig. 6.5. A solution on the new binary tree can be converted to a solution on the original tree easily.

6.4   Optimal CRNF for single sink

The rainbow network flow problem can be greatly simplified if the flow of MDC descriptions from diversity sources is optimized with respect to a given sink. This is particularly so for RNF* since the optimal solution for a single sink does not require duplication of any MDC description.

First, let us reexamine the CRNF problem for a single sink, although optimal CRNF was proven in the previous section to be NP-Hard for multiple sinks. By the definition of CRNF, all MDC descriptions have the same importance and the same rate R, which is treated as the unit rate of 1 for convenience. To the only sink $t \in T$, the optimal

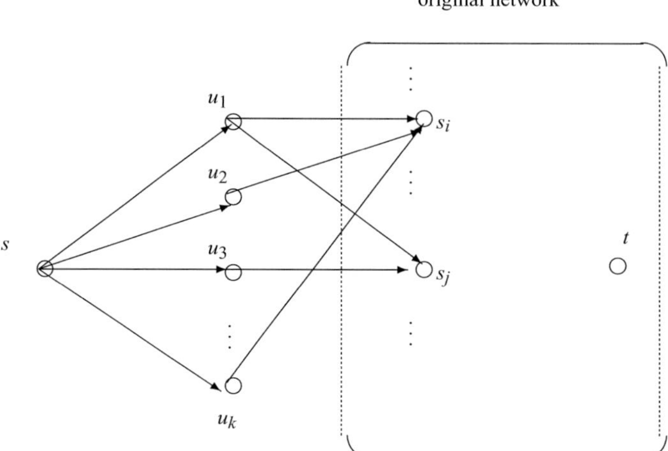

Figure 6.6 Expanded graph with added edges from u_i to the sources that have color i

MDC network flow that achieves the maximum fidelity at the sink t will necessarily have K flow paths of K distinct colors ($K \leq N$), or there is at most one path with respect to each color. These observations allow us to reduce the single-sink CRNF problem to one of conventional maximum network flow, which we call maximum monochrome network flow to distinguish it from maximum-fidelity rainbow network flow.

For each color i, we add a new vertex u_i to the graph G, and add an edge (u_i, s) for each $s \in S$ such that $i \in \Psi_S(s)$. The capacity of edge (u_i, s) is set to 1. Then we add a "super source" s_o, and edges (s_o, u_i) with capacity 1 for each i. Figure 6.6 depicts the resulting expanded graph. We assume that the capacities of all edges in the network are integers. The above construction is valid for arbitrary network topology, and it equates the optimal solution for the single-sink CRNF problem to the maximum monochrome flow from s_o to t in the expanded graph. The latter problem can be solved easily by Goldberg and Tarjan's max-flow algorithm in $O(|E||V|\log(|V|^2/|E|))$ time [43].

The resulting maximum monochrome flow of volume K corresponds to K paths $p_1, \ldots, p_K$ from s_o to t. And each edge e of the graph appears in at most $C(e)$ of the K flow paths. Since the K edges (s_o, u_i), $i = 1, \ldots, K$, are the only outgoing edges of s_o, and each of them has capacity 1, the resulting K flow paths must each go through a different u_i. Referring to Fig. 6.6, if a path p_i from s_o to t first reaches node u_{k_i}, then it must immediately enter a node $s \in S$. Denote by p'_i the remainder subpath of p_i from the node s to t. Clearly, the subpath p'_i is a path from the source s to the sink t of color k_i for the original CRNF problem. Therefore, the set of paths $\{p'_i\}_{i=1,\ldots,K}$, constitutes a solution for the original single-sink CRNF problem. The procedure described above is summarized by the pseudo-code *SingleSink* below.

> **Algorithm SingleSink**
>
> **Input** $G = \langle V, E \rangle$; source set S with spectrum $\Psi_S(s)$, $s \in S$; sink $t \in V$;
> **Output** K paths p'_i, $1 \leq i \leq K \leq N$, each with a different color.
>
> 1. For each color i,
>   add a new vertex u_i to V;
>   for each $s \in S$ such that $i \in \Psi_S(s)$
>   add an edge (u_i, s) with $C(u_i, s) = 1$ to E.
> 2. Add a new vertex s_o to V.
> 3. For each color i,
>   add an edge (s_o, u_i) with capacity $C(s_o, u_i) = 1$ to E.
> 4. Compute the max-flow from s_o to t in the new graph, resulting in
> $p_1, \ldots, p_K$, the K different paths from s_o to t that compose the flow.
> 5. For i from 1 to K,
>   remove the first two edges in p_i to get a new path p'_i.
> 6. Output $p'_1, \ldots, p'_K$.

THEOREM 6.3 *The flow computed by Algorithm SingleSink is an optimal solution of CRNF.*

Proof Clearly, the paths computed by the algorithm form a feasible solution of CRNF. We only need to show that the solution is optimal. This can be proved by contradiction.

Suppose that we have another solution of $K' > K$ paths with different colors $i_1, \ldots, i_{K'}$. Let the K' paths be $q'_1, \ldots, q'_{K'}$. Let q'_i have color k_i and q'_i connect from a source node $s_i \in S$ such that $k_i \in \Psi_S(s_i)$ to the sink t. Then there is an edge (u_{k_i}, s_i) in the expanded graph. As a result, $q_i = (s_o, u_{k_i}) \cdot (u_{k_i}, s_i) \cdot q'_i$ is a path from s_o to t in the expanded graph. The collection of these paths $\{q_i\}_{i=1,\ldots,K'}$ forms a monochrome network flow from s_o to t whose total flow volume is $K' > K$. This is contradictory to the optimality of step 4. $\square$

For CRNF on an undirected graph with a single sink, the above algorithm works as well with virtually no modification. The only concern is that the constructed graph (Fig. 6.6) now has all the newly added edges directed, and all the edges in the original network undirected. However, in the traditional network flow problem, each undirected edge can be converted to two opposite directed edges with the same capacity. The optimal solution of the resulting directed graph will be the same as the optimal solution of the original graph with undirected edges. As a result, Algorithm SingleSink is also an efficient algorithm for CRNF on undirected graphs with single sinks. We have the following theorem.

THEOREM 6.4 *Algorithm SingleSink computes an optimal solution of the single-sink undirected CRNF problem.*

6.5   Concluding remarks

The previous chapter introduced a method for lossy network multicast which relied on efficient multicast of multiple-description encoded packets from the source to the receivers. The problem in its general form was called the Rainbow Network Flow problem, or RNF. In this chapter, we studied the complexity and algorithmic issues for solving the RNF problem, and in particular, a special case variation of it called the CRNF. Unfortunately, the CRNF problem is hard unless for restricted classes of graphs, such as tree topologies. We assumed an integer size for the descriptions. In the next chapter, we will study the theoretical significance of the size of the descriptions, in more detail.

7 Continuous rainbow network flow: rainbow network flow with unbounded delay

In Chapter 5, we introduced a practical approach to the NASCC problem by optimal diversity routing of MDC code streams (the RNF problem) and optimal design of the MDC codes by PET technique. There, we briefly discussed the role of the common rate of the descriptions r and the total number of possible descriptions K. The developments so far assumed communication with bounded delays, in which case the values of r, K become particularly important.

When the delay constraint is relaxed, we find that the set of all achievable distortion tuples converges to a limit independent of the description rate r. This limiting region will turn out to have a simple representation by introduction of a new form of flow we call continuous Rainbow Network Flow, or co RNF, as opposed to the discrete version of the problem considered so far in the thesis. co RNF can be viewed as the generalization of the RNF to fractional flows and is the subject of Sections 7.1 and 7.2.

In di RNF, we assumed the existence of K descriptions of equal rate r. co RNF, in one view, relaxes the constraint on description rates and allows for an arbitrary number of descriptions. Therefore, co RNF contains RNF as a special case. On the converse side, we show that the performance achievable with arbitrary description rates can be achieved, arbitrarily closely, using *any* description rate, provided that the delay constraint is relaxed and the number of descriptions is left unbounded. We formalize the concept of co RNF in Section 7.1. Our main achievability results are proved in Section 7.2.

7.1 Continuous rainbow network flow

In what follows, we introduce co RNF and compare the concepts with those of discrete rainbow network flow (di RNF) introduced in Chapter 5. co RNF can be assumed as a generalization of di RNF in which the descriptions can have arbitrary rates and the total number of descriptions K is not bounded. co RNF, therefore, contains di RNF as a special case. Conversely, we show that each co RNF problem is the limit of a proper sequence of di RNF problems.

co RNF(G, R) is defined with the following elements.

(1) A directed graph $G\langle V, E\rangle$ with a node set V and an edge set E.
(2) Two subsets $S = \{s_1, s_2, \ldots, s_{|S|}\}, T = \{t_1, t_2, t_3, \ldots, t_{|T|}\}$ of V representing the set of source and sink nodes respectively.

(3) A function $R : E \to \mathbb{R}^+$ representing the capacity of each link in E.

(4) A function $\rho : \mathbb{N} \to \mathbb{R}^+$, which assigns description rate $\rho(k)$ to each description $k \in \mathbb{N}$. Function ρ induces a measure μ^ρ on $\mathbb{N}$ as follows: $\mu^\rho(x) \triangleq \sum_{k \in x} \rho(k)$, for all $x \subset \mathbb{N}$.

Definitions of flow paths and spectrums are all parallel to the discrete case introduced in Section 5.4. The main difference is that the set of possible descriptions χ is now the whole $\mathbb{N}$ and descriptions can have arbitrary rates. To distinguish between the discrete and the continuous cases, we drop the overhead dots. For instance, α will indicate a flow, while $\dot{\alpha}$ corresponds to the discrete (or integral) case.

In parallel to the notations in Chapter 5, a continuous flow is indicated by $\alpha(f, W)$ and consists of a set W of flow paths in G, and a coloring function $f : W \to \mathbb{N}$.

$\Phi_E(e, W)$ and $\Phi_V(v, W)$ are, just as in the discrete case, the sets of all flow paths in W that contain the link e or the node v, respectively.

The spectrum of an edge $e \in E$, with respect to flow α, is similarly defined as:

$$\Psi_E(\alpha, e) \triangleq \bigcup_{w \in \Phi_E(e, W)} f(w)$$

Likewise, the spectrum of a node v is defined as:

$$\Psi_V(\alpha, v) \triangleq \bigcup_{w \in \Phi_V(v, W)} f(w)$$

An RNF $\alpha(f, W)$ is said to be admissible with capacity function R, if and only if:

$$\mu^\rho(\Psi_E(\alpha, e)) < R(e) \quad \forall e \in E \tag{7.1}$$

Let $\mathcal{F}(G, R, \rho)$ be the set of all admissible co RNF's in G with capacity function R and description rate assignment ρ. Any RNF $\alpha \in \mathcal{F}(G, R, \rho)$ results in an admissible rainbow flow vector (RFV), $\mathbf{q}(\alpha) = (q_t; t \in T) \in \mathbb{R}^{|T|}$, such that:

$$q_t = \mu^\rho(\Psi_V(\alpha, t)). \tag{7.2}$$

Let $\mathcal{Q}(G, R) \subset \mathbb{R}^{|T|}$ be the set of all admissible RFV's for all assignments of description rates ρ:

$$\mathcal{Q}(G, R) \triangleq \bigcup_\rho \bigcup_{\alpha \in \mathcal{F}(G, R, \rho)} \mathbf{q}(\alpha).$$

7.2   Achievability results

We now repeat the parameterization of the MDC designed through PET, as introduced in Section 5.5.1.

For $diRNF(G, R, r, K, n)$, let the set $\dot{\mathcal{H}}$ be the set of all non-negative discrete vectors $\mathbf{y} = (y_i; i = 1, 2, \ldots, K)$ such that $\sum_{i=1}^{K} y_i = 1$. For future reference, define a subset $\dot{\mathcal{H}}_N$ of all vectors $\mathbf{y}$ in $\dot{\mathcal{H}}$ that have at most N non-zero elements.

For co RNF, on the other hand, let $\mathcal{H}$ consist of all non-negative functions $y : \mathbb{R}^+ \to \mathbb{R}^+$ such that $\int_0^\infty y(a)da = 1$ and that $\int_0^x y(a)da$ is continuous except possibly at a finite number of points. Similarly, define a subset $\mathcal{H}_N$ of all functions y in $\mathcal{H}$ for which $\int_0^x y(a)da$ is discontinuous at most at N points.

In Section 5.5.1, it was shown that there is a one-to-one correspondence between the members of set $\dot{\mathcal{H}}$ and the multiple-description codes generated by the PET technique. Furthermore, the reconstruction distortion at each sink node can be calculated, given the discrete RFV $\dot{\mathbf{q}}$, and any vector $\dot{\mathbf{y}} \in \dot{\mathcal{H}}$ representing the choice of the MDC.

For $diRNF(G, R, r, K)$, define the $|T|$-tupled real valued vector $\dot{\mathbf{d}}(y, \mathbf{q}) = (\dot{d}_t;$ $t \in T) \in \mathbb{R}^{|T|}$ such that:

$$\dot{d}_t(\mathbf{y}, \mathbf{q}) = D_X \left(n^{-1} r \sum_{i=0}^{nq_t/r} i y_i \right). \tag{7.3}$$

Let $\dot{\mathcal{B}}(G, R, r, K, n) \subset \mathbb{R}^{|T|}$ be the union of all such vectors $\dot{\mathbf{d}}$, that is,

$$\dot{\mathcal{B}}(G, R, r, K, n) \triangleq \bigcup_{\dot{y} \in \dot{\mathcal{H}}, q \in \dot{\mathcal{Q}}(G, R, r, K, n)} \dot{\mathbf{d}}(\dot{y}, \mathbf{q}).$$

It should be evident by now that for $diRNF(G, R, r, K, n)$, $\dot{\mathcal{B}}(G, R, r, K, n)$ depends only on the cardinality of the set χ (i.e., K) and not the actual set of descriptions.

For co RNF, the vector $\mathbf{d} = \mathbf{d}(y, \mathbf{q}) = (d_t; t \in T)$ is defined such that:

$$d_t(y, \mathbf{q}) \triangleq D_X \left(\int_0^{q_t} xy(x)dx \right). \tag{7.4}$$

Also, we define $\mathcal{B}(G, R) \subset \mathbb{R}^{|T|}$ to be the union of all $\mathbf{d}(y, \mathbf{q})$ over $y \in \mathcal{H}$ and $\mathbf{q} \in \mathcal{Q}(G, R)$.

Our goal in Section 7.2 will be to prove that the set $\mathcal{B}(G, R)$ is essentially the limit of all distortion tuples achievable by the method introduced in Chapter 5 when considering all possible description rates r and their total number K and removing the delay constraint.

The first step is to establish the achievability result for the discrete version of the problem, a fact we have already considered in Chapter 5 in a less formal setting.

Let $\mathcal{D}_X(G, S, T, R)$ be the set of all achievable distortion $|T|$−tuples, i.e., the set of all tuples $(d_t; t \in T)$ for which a multicast scheme exists with an average distortion of d_t at the sink node $t \in T$. Then:

THEOREM 7.1 $\dot{\mathcal{B}}(G, R, r, K, n) \subset \mathcal{D}_X(G, R, S, T)$ for all $n, K \in \mathbb{N}, r \in \mathbb{R}^+$.

Proof   The achievability is a direct consequence of our construction and was discussed in detail in Chapter 5. In particular, any achievable RFV $\dot{\mathbf{q}} = (\dot{q}_t, i \in T)$, defined in (7.2), indicates that $n\dot{q}_t/r$ distinct descriptions, each of rate r, can be communicated to sink nodes $t \in T$. To find the reconstruction distortion at a node $t \in T$ using an arbitrary MDC, we would need to know the exact subset of χ that is present at t. However, $n\dot{q}_t/r$ only specifies the total *number* of the distinct descriptions received and not the exact subset of the descriptions. That is why we need to assume that the MDC is *balanced*, i.e., for any $0 \leq k \leq K$, the source can be reconstructed to the same fidelity given any subset of size k out of the total of K descriptions. In this case, knowing the total number of *distinct* descriptions available at $t \in T$ (i.e., $n\dot{q}_t/r$) suffices to find the reconstruction distortion at t.

For producing balanced MDC, we have chosen to use the Priority Encoding Transmission (PET) technique, with which any number of balanced multiple descriptions can be produced from a progressively encoded source stream. From Section 5.5.1, any non-negative vector $\dot{\mathbf{y}} = (y_1, y_2, \ldots, y_K)$, such that $\sum_{i=1}^{K} y_i = 1$, specifies a valid PET. Theorem 7.1 is now easily proved by comparing (5.8) and (7.3). $\qquad\square$

The following simple theorem establishes the fact that the distortion tuples achievable by di RNF are subsets of $\mathcal{B}(G, R)$.

THEOREM 7.2   *For all r, K, n,*

(1) $\dot{\mathcal{Q}}(G, R, r, K, n) \subset \mathcal{Q}(G, R)$,
(2) $\dot{\mathcal{B}}(G, R, r, K, n) \subset \mathcal{B}(G, R)$

Proof   Evidently, di RNF flows are special cases of co RNF flows in which all descriptions have the same rate, i.e., $\rho(k)$ is equal to r for exactly K descriptions and is zero for the rest. This establishes the first part of the theorem. To prove the second part, take any $\dot{\mathbf{y}} \in \dot{\mathcal{H}}$. Then, define $\mathbf{y}(x) = \sum_{l=1}^{K} \delta^{(Di)}(x - l \cdot r)$ where $\delta^{(Di)}(\cdot)$ is Dirac's delta function. For the same flows, and with this choice of the function y, the two distortion vectors in Equations (7.3) and (7.4) will be equal, which proves the second part of the theorem. $\qquad\square$

A set $A \subset \mathbb{R}^N$ is said to be dense in another set $B \subset \mathbb{R}^N$, if, (1) $A \subset B$ and (2) for any $b \in B$ and any $\epsilon > 0$, there exists a $b \in B$ such that $||a - b|| \leq \epsilon$, where $|| \cdot ||$ is some metric on $\mathbb{R}^N$.

Let R_S be the sum of the capacity of all links that flow out of the set of server nodes S. The following are our main theorems.

THEOREM 7.3   *Let $\mathbf{r} = r_1, r_2, \ldots$ be any sequence of positive real numbers that converges to zero. Then for all integer delays n,*

$$\dot{\mathcal{B}}(\mathbf{r}) \triangleq \bigcup_m \dot{\mathcal{B}}(G, R, r_m, \lceil R_S/r_m \rceil, n)$$

is dense in $\mathcal{B}(G, R)$.

In other words, points in $\mathcal{B}(G, R)$ can be approximated arbitrarily closely using points achievable by a discrete flow with small enough description rate. The above theorem, with the aid of some lemmas, will result in the following.

THEOREM 7.4   *For all $r > 0$,*

$$\tilde{\dot{\mathcal{B}}}(r) \triangleq \bigcup_{n \in \mathbf{N}} \dot{\mathcal{B}}(G, r, \lceil (R_S/r) \cdot n \rceil, n) \text{ is dense in } \mathcal{B}(G, R).$$

Therefore, points in $\mathcal{B}(G, R)$ can be approximated arbitrarily closely using points achievable by discrete flows with any description rates for large enough delay values.

Either of these two theorems, together with the fact that $\dot{\mathcal{B}}(G, R, r, K, n) \subset \mathcal{D}_X$ (G, S, T, R) for all n, K and r (see Theorem 7.1), result in the following achievability result about the points in set $\mathcal{B}(G, R)$:

THEOREM 7.5   *For any $\epsilon > 0$ and any $\mathbf{d} \in \mathcal{B}(G, R)$, one has:*

$$\mathbf{d} + \epsilon \mathbf{1}_{|T|} \in \mathcal{D}_X(G, S, T, R)$$

where $\mathbf{1}_N$ is a vector of all ones of length N.

To summarize, Theorem 7.2 establishes the fact that solutions to the NASCC problem achievable with di RNF are subsets of the co RNF region. Conversely, Theorems 7.3 and 7.4 show that the co RNF region can be approximated arbitrarily closely using a sequence of di RNF regions when either the description rates go to zero or the delays go to infinity. These theorems will then result in Theorem 7.5, which establishes the fact that the co RNF region is indeed a valid solution to the NASCC problem.

To prove Theorems 7.3 and 7.4, we need to establish a series of theorems and lemmas and in particular consider the role of communication delay and description rates in much more detail.

The following theorem, proved later in this section, establishes the fact that the set of distortion tuples achievable by using any series of description rates that goes to zero, is essentially independent of the series itself.

THEOREM 7.6   *Let $\mathbf{r} = r_1, r_2, \ldots,$ $\mathbf{u} = u_1, u_2, \ldots$ be any two sequences of positive real numbers that converge to zero. Then, $\dot{\mathcal{B}}(\mathbf{r}), \dot{\mathcal{B}}(\mathbf{u})$, are both dense in*

$$\dot{\mathcal{B}}(\mathbf{r}) \cup \dot{\mathcal{B}}(\mathbf{u})$$

where $\dot{\mathcal{B}}(\mathbf{r}), \dot{\mathcal{B}}(\mathbf{u})$ are defined as in Theorem 7.3.

Roughly speaking, Theorems 7.6 and 7.3 prove the following facts: (a) distortion tuples achievable through co RNF are achievable with di RNF by using descriptions of small rates (Theorem 7.3): and (b) which sequence of description rates is used to approximate co RNF is irrelevant, as they all result in essentially the same achievable regions (Theorem 7.6).

Likewise, the distortion tuples achievable by considering all delays is, in essence, independent of the description rate as formalized in the following theorem.

THEOREM 7.7   *For all r, r', both $\widetilde{\mathcal{B}}(r)$ and $\widetilde{\mathcal{B}}(u)$ are dense in $\widetilde{\mathcal{B}}(r) \cup \widetilde{\mathcal{B}}(u)$ where $\widetilde{\mathcal{B}}(\mathbf{r}), \widetilde{\mathcal{B}}(\mathbf{u})$ are defined as in Theorem 7.4.*

The above theorem states that achievable regions when the delay constraint is relaxed are essentially the same regardless of the rate of the descriptions used in communication. Theorem 7.4, on the other hand, shows that any of these regions is dense in that of co RNF.

We start by analyzing the achievable region of di RNF in more detail. In particular, we examine the role of description rates and their total number under more scrutiny. First, we show that increasing the number of descriptions can only increase the achievable region. Therefore, in di RNF, there is no harm in letting the set χ be the whole integers $\mathbb{N}$ (i.e., $K = \infty$).

LEMMA 7.1   *For all $r \in \mathbb{R}^+$ if $K > K' \in \mathbb{N}$ then:*

$$\dot{\mathcal{B}}(G,R,r,K',n) \subset \dot{\mathcal{B}}(G,R,r,K,n) \subset \dot{\mathcal{B}}(G,R,r,\infty,n).$$

Proof   Take $\dot{\mathbf{d}}$ any member of $\dot{\mathcal{B}}(G,R,r,K,n)$ which corresponds to some $\mathbf{y} \in \dot{\mathcal{H}}$ and a flow vector $\dot{\mathbf{q}} \in \dot{\mathcal{Q}}(G,R,r,K,n)$. Trivially, for $K' > K$, the flow vector $\dot{\mathbf{q}}$ remains achievable (one simply ignores all the other $K' - K$ possible descriptions).

Define $\dot{\mathbf{y}}'$ such that $y'_l = y_l, l = 1,2,\ldots,K$ and $y'_l = 0, l = K+1,\ldots,K'$. Let the distortion vector $\dot{\mathbf{d}}' \in \dot{\mathcal{B}}(G,R,r,K',n)$ correspond to $\dot{\mathbf{y}}'$ and the flow vector $\mathbf{q}'$. Obviously, $\dot{\mathbf{d}}'$ is equal to $\dot{\mathbf{d}}$, proving the assertion. $\square$

There is, therefore, no loss in the performance if one assumes that the number of descriptions K is infinity. It should be noted, however, that at most $K_S(n,r) = \lceil n \cdot R_S/r \rceil$ descriptions can flow out of all the server nodes, where R_S is the sum of the outgoing capacity of all the server nodes in S. Therefore, one can safely limit the number of descriptions to $K_S(n,r)$.

LEMMA 7.2   *For all $r > 0$ and any $n, K \in \mathbb{N}$,*

(1) $\dot{\mathcal{Q}}(G,R,r,K,n) = \dot{\mathcal{Q}}(G,R,r/n,K,1)$
(2) $\dot{\mathcal{B}}(G,R,r,K,n) = \dot{\mathcal{B}}(G,R,r/n,K,1)$

In other words, a problem with delay n can be replaced with another problem with delay one, in which the description rates are all divided by n. The proof of the lemma is based on a simple time-sharing argument. It is also obvious from Eq. (5.9). Note that the product $n^{-1}r$ always appears together. Therefore, one can interpret a flow with delay n and rate r as a flow with delay one and rate r/n.

The following lemma shows that delay improves the performance when the number of descriptions is scaled appropriately.

LEMMA 7.3   *For any $n, K \in \mathbb{N}$:*

(1) $\dot{\mathcal{Q}}(G,R,r,K,1) \subset \dot{\mathcal{Q}}(G,R,r,n \cdot K,n)$
(2) $\dot{\mathcal{B}}(G,R,r,K,1) \subset \dot{\mathcal{B}}(G,R,r,n \cdot K,n)$

Proof   From Lemma (7.2) it suffices to show that

$$\dot{\mathcal{Q}}(G,R,r,K,1) \subset \dot{\mathcal{Q}}(G,R,r/n,n \cdot K,1)$$

and similarly for $\dot{\mathcal{B}}$.

To show this, note that any flow path w can be split into n flows, each of rate r/n and keeping all admissible flows still admissible. Now, expand the set χ to a set χ' such that $|\chi'| = i|\chi| = iK$. To each member of χ, we can therefore assign i distinct members of χ'. For any admissible flow vector $\dot{\mathbf{q}} \in \mathcal{Q}(G,R,r,K,1)$ and for any integer n, there exists an admissible flow vector $\dot{\mathbf{q}}' \in \mathcal{Q}(G,R,r/n,n \cdot K,1)$ such that $\dot{q}'_t = \dot{q}_t$, which proves the first parts of the lemma.

Now let $\dot{\mathbf{d}}$ be a distortion vector corresponding to $\dot{\mathbf{q}}$ and $\dot{\mathbf{y}}$. Then define $\dot{\mathbf{y}}'$, such that $y'_{ik} = y_k$ for all $k = 1,2,\ldots,K$, and let $y'_{k'} = 0$ for all k' not divisible by n. The distortion vector $\dot{\mathbf{d}}'$ corresponding to $\dot{\mathbf{y}}'$ and $\dot{\mathbf{q}}$ is thus equal to $\dot{\mathbf{d}}$. Therefore, any distortion vector

$\mathbf{\dot{d}'} \in \dot{\mathcal{B}}(G, R, r, K, 1)$ is equal to a distortion vector $\mathbf{\dot{d}}$ in $\dot{\mathcal{B}}(G, R, r/n, R, n \cdot K, 1)$ which proves the assertion. $\qquad\square$

From the above lemma, for any r, the largest achievable region will occur when the delay n goes to infinity, as expected, or when the size of the descriptions r/n goes to zero. This, however, does not necessarily mean that the limits $\lim_{r\to 0^+} \dot{\mathcal{Q}}(G, R, r, K(r), 1)$ or $\lim_{n\to\infty} \dot{\mathcal{Q}}(G, R, r, K(n), n)$ or their corresponding counterparts in $\dot{\mathcal{B}}$ exist (note that we are recognizing the possible dependence of the number of descriptions, K, to the delay or description rates).

For instance, $\dot{\mathcal{Q}}(G, R, r, K, 1)$ is not necessarily continuous for all $r > 0$. Take for example the extreme case where the size of the description r is equal to the maximum capacity of outgoing links of all server nodes. Then, no description packet is able to leave any source. When r is slightly decreased, however, it might be possible to have a single description flow and therefore the set of achievable RFV's can jump discontinuously.

As the description size becomes smaller (or when the delay become larger), however, this discontinuity becomes less and less significant. The next two lemmas formalize this idea.

So far, we have considered vectors $\mathbf{\dot{y}} \in \dot{\mathcal{H}}$. The set $\dot{\mathcal{H}}$ is convex and is hence suitable for optimization purposes carried out in Chapter 5. We now show that one can confine derivations to set $\dot{\mathcal{H}}_{|T|} \subset \dot{\mathcal{H}}$ without loss of generality. It should be noted, however, that the set $\dot{\mathcal{H}}_{|T|}$ is not convex anymore. This observation is crucial for the proof of Lemma 7.5, presented later in this section.

We start with the following lemma.

LEMMA 7.4 *Take any $\mathbf{\dot{q}} = (\dot{q}_t; t \in T) \in \dot{\mathcal{Q}}(G, R, r, K, n)$ and $\mathbf{\dot{y}} = (y_1, y_2, \ldots, y_K) \in \dot{\mathcal{H}}$. Then, there exists a $\mathbf{\dot{y}'} \in \dot{\mathcal{H}}_{|T|}$ such that $\dot{d}_t(\mathbf{\dot{y}'}, \mathbf{\dot{q}}) \leq \dot{d}_t(\mathbf{\dot{y}}, \mathbf{\dot{q}})$ for all $t \in T$, where $\dot{d}_t s$ are defined as in Eq. (7.3).*

Proof Collect equal elements of the vector $\mathbf{\dot{q}} = (\dot{q}_t; t \in T)$ together. More precisely, let $Z(c) = \{t : n\dot{q}_t/r = c\}$. Since there are $|T|$ sink nodes only, at most a number $|T|$ of the sets $Z(1), Z(2), \ldots, Z(K)$ are non-empty. In other words, there exist a sequence $\kappa = (k_1, k_2, \ldots, k_{|T|})$ such that $Z(c)$ can be non-empty only when $c \in \kappa$. Let's assume $k_1 < k_2 < \cdots < k_{|T|}$. Construct a vector $\mathbf{\dot{y}'} = (y'_1, y'_2, \ldots, y'_K)$ as follows. Initially, set $y'_i = 0$ for all $i = 1, 2, \ldots, K$. Then, for each $k_i \in \kappa$ let $y'_{k_i} = \sum_{j=k_{i-1}+1}^{k_i} y_j$, where $k_0 = 1$ is assumed for convenience. Evidently, the vector $(y'_1, y'_2, \ldots, y'_K)$ has at most $|T|$ non-zero elements and the sum of all its elements is 1. Therefore, $\mathbf{\dot{y}'} \in \dot{\mathcal{H}}_{|T|}$.

To prove the lemma, it suffices to show that $\sum_{i=0}^{k} iy_i \leq \sum_{i=0}^{k} iy'_i$ for all $k_j \in \kappa$. For $k_j \in \kappa$ we have

$$\sum_{i=0}^{k_j} iy_i = \sum_{i=1}^{k_1} iy_i + \sum_{i=k_1+1}^{k_2} iy_i + \cdots + \sum_{i=k_{j-1}+1}^{k_j} iy_i$$

$$\leq k_1 \sum_{i=1}^{k_1} y_i + k_2 \sum_{i=k_1+1}^{k_2} y_i + \cdots + k_j \sum_{i=k_{j-1}+1}^{k_j} y_i$$

$$= k_1 y_1' + k_2 \ddot{y}_2' + \cdots + k_j y_j'$$

$$= \sum_{i=0}^{k_j} i y_i'. \qquad \square$$

This lemma is necessary to prove the following lemma.

LEMMA 7.5 *Take any $r, r' > 0$, a $\dot{\mathbf{q}}' \in \dot{\mathcal{Q}}(G, R, r', K', 1)$ and a $\dot{\mathbf{d}}' \in \dot{\mathcal{B}}(G, R, r', K', 1)$.*

If $D_X(\cdot)$ is differentiable everywhere with bounded derivatives (which should be the case at least for all stationary signals), then, for any

$$K \geq K' \cdot (\max\{r, r'\} / \min\{r, r'\} + 1)$$

there exists a $\dot{\mathbf{q}} \in \dot{\mathcal{Q}}(G, R, r, K, 1)$ and a $\dot{\mathbf{d}} \in \dot{\mathcal{B}}(G, R, r, K, 1)$, such that $\forall t \in T$:

(1) $\left| \dot{q}_t - \dot{q}_t' \right| \leq |E|^2 r$

(2) $\left| \dot{d}_t - \dot{d}_t' \right| \leq \gamma_q |T| \cdot |E|^2 r + \gamma_q' (|T| \cdot |E|^2 r)^2$ *for some constants $\gamma_q, \gamma_q' > 0$ depending only on the rate-distortion function $D_X(\cdot)$.*

Proof **Case 1**, $r > r'$: Let's fist prove the lemma assuming that $r > r'$, which is the somewhat more involved case. Let $\dot{\mathbf{q}}'(\dot{\alpha}') \in \dot{\mathcal{Q}}(G, R, r', K, 1)$. If the size of the descriptions were increased from r' to $r > r'$, the volume of the flow at every edge e (see Eq. 7.2) increases to $|\dot{\Psi}_E(\dot{\alpha}', e)| \times r$. This of course might render the flow infeasible because some of the conditions in (7.1) might not be satisfied any more. For any such overloaded edge, we delete some of the flows randomly until the capacity constraint on the edge is respected.

We know that a total spectrum of $|\dot{\Psi}_E(\dot{\alpha}', e)| \times r'$ using descriptions of rate r' can flow through e. When we replace these descriptions with descriptions of rate $r > r'$, we can keep at least $\lfloor r' |\dot{\Psi}_E(\alpha, e)| / r \rfloor$ of these flow paths that correspond to *distinct* descriptions. The total spectrum of e therefore is at least $(r' |\dot{\Psi}_E(\alpha, e)| / r - 1) \times r = r' |\dot{\Psi}(\alpha, e)| - r$. In other words, there is a flow with descriptions of rate r that respects the capacity on the edge e and does not decrease the total spectrum of e by more than r.

Now replacing flows of rate r' with flows of rate $r > r'$ for edge e might also disrespect the constraint on other edges. For each such edge, however, we can repeat the procedure, i.e., replacing all the flows with rate r' with flows of rate r and then deleting extra flows until the capacity on that edge is respected. Since there are at most $|E|$ edges, after replacing all the flows with flows of rate r and deleting excess flows, the total reduction in the overall flow is at most $r \times |E|^2$. Therefore, for each sink node $t \in T$, a flow of at least $\dot{q}_t = \dot{q}_t' - r|E|^2$ is still feasible, which leads to the first part of the lemma.

Now for any choice of the PET code using a vector $\dot{\mathbf{y}} \in \mathcal{H}_{|T|}$, a distortion vector $\dot{\mathbf{d}}' = (\dot{d}_t'; t \in T)$ is obtained. Note that due to Lemma 7.4, we have confined ourself to vectors in set $\mathcal{H}_{|T|}$ without loss of generality.

Replacing the old flow with this new flow, a new distortion vector $\dot{\mathbf{d}} = (\dot{d}_t; t \in T)$ can be achieved such that

$$\dot{d}_t = D_X \left(r' \sum_{k=1}^{\dot{q}_t/r'} k y_i - r\varsigma \right).$$

The value of ς is bounded as follows. Since $\dot{\mathbf{y}} \in \dot{\mathcal{H}}_{|T|}$, there are at most $|T|$ non-zero values of y_i for $i = 1, 2, \ldots, K$. For each such value, there is at most a decrease of $r|E|^2$ in the summation. Therefore, $\varsigma \le |T| \cdot |E|^2$.

Thus:

$$\dot{d}_t = D_X \left(r' \sum_{k=1}^{\dot{q}_t/r'} k y_i \right) + Z_X \varsigma r + o(Z_X \varsigma r)$$

$$= \dot{d}_t + Z_X \varsigma r + o(Z_X \varsigma r)$$

where we have used the fact that $D_X(\cdot)$ is differentiable with a bounded derivative and Z_X is a constant depending only on the function D_X.

Therefore,

$$|\dot{d}_t - d_t| \le |T| \cdot |E|^2 Z_X r + o(Z_X |T| \cdot |E|^2 r)$$

which proves the result.

Case 2, $r < r'$: Take n, any integer such that $r'/n < r$. Since $\dot{\mathcal{Q}}(G, R, r', K, 1) \subset \dot{\mathcal{Q}}(G, R, r'/n, n \cdot K, 1)$, it suffices to prove the statement of the lemma for all $\dot{q}' \in \dot{\mathcal{Q}}(G, R, r'/n, n \cdot K, 1)$. But, since $r > r'/n$, from Case 1 of the proof, there exists a $\dot{\mathbf{q}} \in \dot{\mathcal{Q}}(G, R, r', K, 1)$ such that

$$|\dot{q}_t - \dot{q}'_t| \le |E|^2 \cdot r \quad \forall t \in T$$

which proves the result immediately. The statement on $\dot{\mathcal{B}}$ follows similarly. Note that we only need $K' \le n \cdot K \le (r'/r)K$ descriptions. $\qquad\square$

The following proposition can now be proved. It states that in fact any point achievable using a given description rate is also closely achievable using any other description rate, with a high enough delay.

PROPOSITION 7.1   *Take any* $r, r', \epsilon > 0$. *For every* $\dot{\mathbf{q}} \in \dot{\mathcal{Q}}(G, R, r, K, n)$, $\dot{\mathbf{d}} \in \mathcal{B}(G, R, r, K, n)$ *there exist* n', $\dot{\mathbf{q}}' \in \dot{\mathcal{Q}}(G, R, r', nn' \cdot K, nn')$ *and* $\dot{\mathbf{d}}' \in \mathcal{B}(G, R, r', nn' \cdot K, nn')$, *such that*

(1) $|\dot{q}_t - \dot{q}'_t| \le \epsilon$ *and* (2) $|\dot{d}_t - \dot{d}'_t| \le \epsilon$ *for all* $t \in T$.

Proof   To prove the first part, take

$$n' = \max\{r' \lceil |E|^2/\epsilon \rceil, \lceil r'\gamma_q |T| \cdot |E|^2/\epsilon \rceil, r \lceil |E|^2/\epsilon \rceil, \lceil r\gamma_q |T| \cdot |E|^2/\epsilon \rceil\}$$

where γ_q is the constant in Lemma 7.5.

Since

$$\dot{\mathcal{Q}}(G, R, r, K, n) = \dot{\mathcal{Q}}(G, R, r/n, K, 1) \subset \dot{\mathcal{Q}}(G, R, r/(nn'), K \cdot nn', 1),$$

it suffices to prove part (1) assuming that $\dot{\mathbf{q}} \in \dot{\mathcal{Q}}(G, R, r/(nn'), nn' \cdot K, 1)$.

But from Lemma (7.5), for every $\dot{\mathbf{q}} \in \mathcal{Q}(G, R, r/(nn'), nn' \cdot K, 1)$, there should exist a $\dot{\mathbf{q}}' \in \mathcal{Q}(G, R, r'/(nn'), nn' \cdot K, 1)$ such that

$$\left| \dot{q}_t - \dot{q}'_t \right| \leq \max\{(r'/(nn')), r/(nn')\} \times |E|^2 \leq \epsilon \quad \forall t \in T,$$

which proves the first part of the proposition. The second part follows similarly. $\qquad\square$

We are now in a position to prove Theorem (7.6).

Proof of Theorem 7.6   Take the two sequences $\mathbf{r} = \{r_1, r_2, \ldots\}$ and $\mathbf{u} = \{u_1, u_2, \ldots\}$. It suffices to show that for all $\dot{\mathbf{d}} \in \dot{\mathcal{B}}(\mathbf{r})$, there exists a $\dot{\mathbf{d}}' \in \dot{\mathcal{B}}(\mathbf{u})$ such that

$$\left| \dot{d}_t - \dot{d}'_t \right| \leq \gamma_d \epsilon$$

for some constant γ_d.

To show this, let $\dot{\mathbf{q}} \in \dot{\mathcal{B}}(G, R, r_i, K, 1)$ for some K. Now take $\dot{\mathbf{q}}' \in \dot{\mathcal{B}}(G, R, u_i, K, 1)$ for some $u_j \leq r_i/(R_S Z_X |T| \cdot |E|^2)$. One such u_j should exist, since the sequence $\mathbf{u}$ converges to zero. The statement of the theorem follows from Lemma 7.5. $\qquad\square$

Proof of Theorem 7.7   This theorem follows from Theorem 7.6 by letting $r_i = r/i$ and $u_i = r'/i$. $\qquad\square$

Finally, to prove Theorems 7.3 and 7.4, we need to show that any co RNF can be approximated arbitrarily closely by a di RNF with a sufficiently large delay (or equivalently, with a sufficiently small description rate).

PROPOSITION 7.2   *Let $\mathbf{q} \in \mathcal{Q}(G, R)$, $\mathbf{d} \in \mathcal{B}(G, R)$ and $\epsilon, \epsilon' > 0$ be given. Then, there exists an $r > 0$, an integer K, a $\dot{\mathbf{q}} \in \dot{\mathcal{Q}}(G, R, r, K, 1)$, and $\dot{\mathbf{d}} \in \dot{\mathcal{B}}(G, R, r, K, 1)$ such that*

$$|q_t - \dot{q}_t| \leq \epsilon \tag{7.5}$$

and

$$|d_t - \dot{d}_t| \leq \epsilon' \tag{7.6}$$

for all $t \in T$.

Proof   Take any $\mathbf{q} \in \mathcal{Q}(G, R)$. This RVF corresponds to some assignment of description rates ρ and an admissible flow $\alpha \in \mathcal{F}(G, R, \rho)$. Let v be the set of all descriptions with non-zero rate, i.e., $v = \{k : \rho(k) > 0\}$. Let $r_{min} = \min_{k \in v} \rho(k)$. Take $r^* = \epsilon r_{min}/|E|^2$. With the same arguments as in the proof of Lemma 7.5, we can show that the flow of each description k with rate $\rho(k)$ can be replaced with $\lfloor \rho(k)/r^* \rfloor$ descriptions of *equal* rates r^* to obtain a $\dot{\mathbf{q}}^* \in \dot{\mathcal{Q}}(G, R, r^*, K_S(r^*), 1)$ such that for all $t \in T$ we have $\left| \dot{q}^*_t - q_t \right| \leq \epsilon$.

To prove (7.6), we also need the additional fact that for every $\mathbf{y} \in \mathcal{H}$, $\int_0^x y(a)da$ and therefore, $\int_0^x ay(a)da$ is continuous except possibly at a finite set of points. Also, we need the equivalent of Lemma 7.4 to restrict ourselves to $\mathcal{H}_{|T|} \subset \mathcal{H}$. We can then show that $\int_0^{q_t} xy(x)dx$ can be approximated arbitrarily closely by a Riemann integral using intervals of length r^* by choosing r^* small enough. The details of the proof are omitted. $\qquad\square$

We can now prove Theorems 7.3 and 7.4.

Proof of Theorems 7.3 and 7.4 Take a $\mathbf{q} \in \mathcal{Q}(G, R)$ and take r, K to satisfy the requirements in Proposition 7.2 for $\epsilon = \epsilon' = \varepsilon/2$. Now take $n = \max\{\lceil |E|^2/\epsilon \rceil, \lceil \gamma_q |T| \cdot |E|^2/\epsilon \rceil\}$. Therefore, each point in $\mathcal{Q}(G, R)$ is closer than $\varepsilon/2$ to some point in $\dot{\mathcal{Q}}(G, R, r, K, n)$. The same argument applies to points in $\mathcal{B}(G, R)$, which proves the theorem. $\qquad\square$

Theorem 7.5 follows directly from Theorems 7.3 and 7.4.

7.3   Concluding remarks

The focus of this chapter was to derive the limit of the achievable performance with methods introduced in Chapter 5. The chapter answered many questions regarding the role of description rates (r), the number of descriptions (K), and the delay (n). To sum, the quality of the optimal designs will not decrease if one (a) increases K, or (b) increases n, or (c) takes the description rates r to be very small.

We found the limit of achievable distortion tuples over all possible delays, rate of descriptions, and their total number. The result was shown to have a simple representation by introducing a new form of flow we called Continuous Rainbow Network Flow (co RNF). In one view, co RNF is the generalization of RNF to fractional flows. The "operational" significance of this fractional generalization is established through Theorem 7.5.

8   Practical methods for MDC design

Multiple-description codes are powerful tools for network-aware source coding and communication, as suggested in the previous chapters. We showed how MDC can be useful even in error-free networks. MDC, however, traditionally has been used to combat losses in packet lossy networks in which packets are likely to be dropped or lost. In this chapter, we review practical techniques for construction and optimization of MDCs.

In the most general setting, an MDC scheme generating K descriptions can be regarded as a system consisting of K encoders (also called side encoders), and $2^K - 1$ decoders, one for each subset of descriptions. Figure 8.1 illustrates the block diagram of an MDC scheme for three descriptions. Each encoder generates a bit stream (description) of the same source and sends it to the receiver(s). The sender does not know how many streams are received by a particular receiver, but each receiver has this information. If only some descriptions arrive at a given destination, the decoder corresponding to that subset of descriptions is used to jointly decode them. The K decoders corresponding to individual descriptions are called side decoders, while the others are termed joint decoders. Moreover, the joint decoder corresponding to the whole set of descriptions is known as the central decoder.

8.1   Overview of MDC techniques

Practical MD coding schemes for memoryless sources have been extensively investigated. Some of the most representative approaches are PET-based MDC, MD quantization, and MD correlating transforms. This section offers a brief overview of these three approaches. The chapter continues with details about optimal PET-based and quantization-based MDC design.

PET-based MDC. We briefly reviewed the construction of balanced MDC using the PET technique in Section 5.5.1. This method is used to create symmetric or balanced multiple descriptions and is very popular in image and video transmission applications. It consists of combining a successively refinable (i.e., progressively refinable or scalable, or embedded) source code with uneven erasure protection. The scalable code stream is split into consecutive segments of decreasing importance. Then the source segments with decreasing importance are encoded with progressively weaker erasure protection channel codes. Further, the descriptions are formed across the channel codewords. This

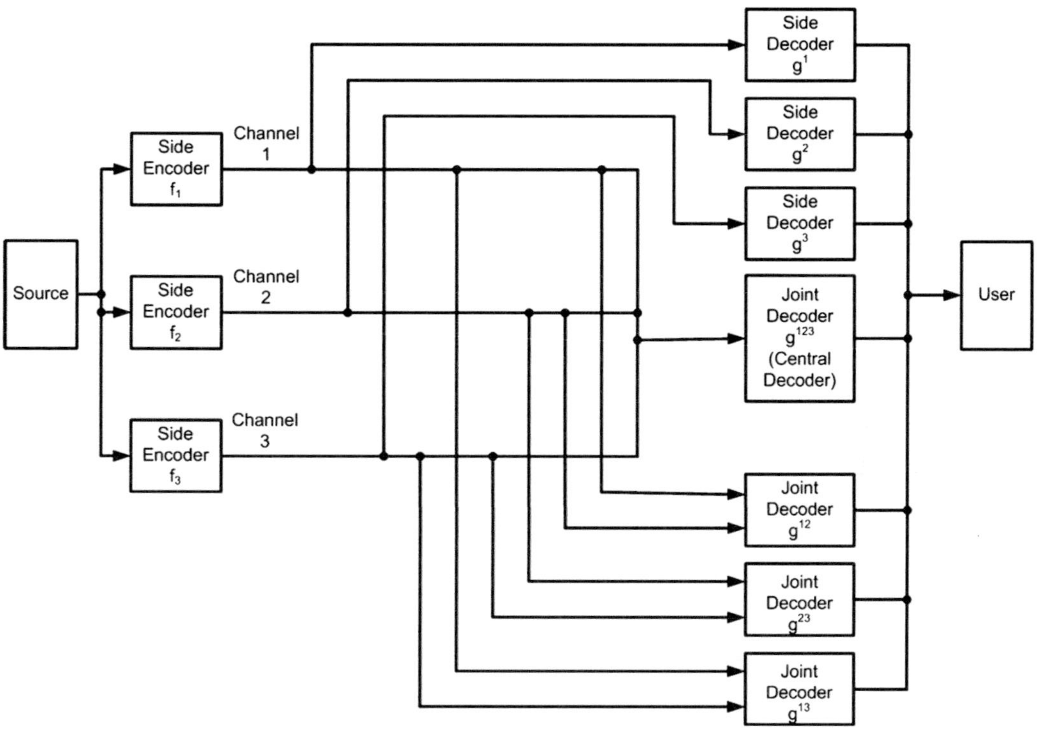

Figure 8.1 Multiple-description scheme for three descriptions

framework was introduced by [72]. The problem of optimal redundancy allocation in rate-distortion sense was addressed in [92], [89], [38], [95], [76], [79].

Although the PET-based MDC performance does not approach the optimal achievable MDC performance arbitrarily closely, it is an attractive approach due to its simplicity and compatibility with scalable image and video coding standards. Moreover, it was recently shown in [96] that regardless of the number of descriptions, the gap between the PET-based MDC scheme and the optimal MD coding can be bounded by a small constant value (precisely, it is less than 1.5 bits in individual description rate).

An improvement to PET-based MDC was proposed in [141], [142], for the case when the scalable code stream can be split into independently decodable sub-streams. This scheme applies interleaved (or permuted) Reed Solomon codes of decreasing strengths across the sub-streams and forms the descriptions across the channel codewords. In [141] the asymptotical performance of this scheme is analyzed for a Gaussian memoryless source, as the rate and code block length approach ∞, while the authors of [142] address the problem of optimal redundancy allocation in the rate-distortion sense, in practical situations.

MD quantization. In MD quantization, the descriptions are generated using separate quantizers. Vaishampayan pioneered the optimal design for MD scalar quantizers (MDSQ) in [44, 98]. He introduced the critical notion of index assignment (IA), which governs the tradeoff between the central and side descriptions, and proposed good index assignments for the case of two balanced descriptions. The design of MD vector quantizers (MDVQ), respectively optimization of IA was addressed in [85,128]. MDVQ is able to approach theoretical limits as the vector dimension increases; however, its complexity increases exponentially with the dimension. Therefore, MD lattice VQ was proposed as a low-complexity MDVQ technique [75,91,99,132].

MD correlating transforms. In this framework the descriptions are generated by a linear transform with the purpose of correlating signals. Redundancy is added through the way transform coefficients are generated [116]. The statistical dependency between transform coefficients can be used to improve the estimation of the lost coefficients. The general approach is to multiply a transform matrix T to the vector of random variables. Let us consider an example as is shown in [116]. Assume X_1, X_2 are two independent, zero-mean Gaussian random variables with variances σ_1^2, σ_2^2, respectively, $\sigma_1^2 \neq \sigma_2^2$. We can form the descriptions (Y_1, Y_2) of (X_1, X_2) by

$$\begin{bmatrix} y_1 \\ y_2 \end{bmatrix} = \begin{bmatrix} \frac{1}{\sqrt{2}} & \frac{1}{\sqrt{2}} \\ \frac{1}{\sqrt{2}} & -\frac{1}{\sqrt{2}} \end{bmatrix} \begin{bmatrix} x_1 \\ x_2 \end{bmatrix}. \tag{8.1}$$

This way, $Y_1 = (X_1 + X_2)/\sqrt{2}$, $Y_2 = (X_1 - X_2)/\sqrt{2}$, and Y_1, Y_2 are correlated with correlation coefficient

$$E[Y_1 Y_2] = \frac{\sigma_1^2 - \sigma_2^2}{\sigma_1^2 + \sigma_2^2}. \tag{8.2}$$

If one description is not present, we can estimate the lost description to better quality than if X_1 and X_2 are directly used as descriptions. [117,118] extend this simple technique to more general transforms and longer vectors.

MD coding with frames. The idea of quantized frame expansion [119, 120] is to left-multiply a source vector $x^{(N)}$ with length N by a matrix $F \in \mathbb{R}^{M \times N}, M > N$ to produce a vector of M transform coefficients. These coefficients are scalar quantized and partitioned into K sets, $K \leq M$, to form K descriptions. The quantized expansion coefficients are represented by $y = Q(Fx)$. The source vector can be reconstructed by solving a least-squares problem: $\arg\min_x \|y - Fx\|^2$. Improvement can be made by introducing more complicated reconstruction methods [121–123].

8.2   Optimal design of multiple-description scalar quantizers (MDSQ)

8.2.1   MDSQ–definition and notations

A K-description MDSQ consists of K side encoders $f_1, \cdots, f_K$, and $2^K - 1$ decoders, each decoder $g_{\mathcal{L}}$ corresponding to a non-empty subset $\mathcal{L}$ of descriptions, $\mathcal{L} \subseteq \mathcal{K} = \{1, \cdots, K\}$. Recall that we use the term central decoder to describe the decoder corresponding to the whole set of descriptions $\mathcal{K}$, and use the term side decoders to describe the decoders corresponding to the sets which have only one description. The decoders other than the side decoders are referred to as joint decoders. Therefore, among the $2^K - 1$

component decoders of an MDSQ, there are K side decoders, one central decoder, and $2^K - 1 - K$ joint decoders. Note that the central decoder is a special case of a joint decoder.

Given a source sample x, first the side encoders map x into some K-tuple $(i_1, \cdots, i_K) \in \mathcal{I}_{\mathcal{K}}$, where i_k is generated by the k-th encoder, $1 \leq k \leq K$, and $\mathcal{I}_{\mathcal{K}}$ denotes the set of all K-tuples generated by all side encoders. Formally, the encoder f_k is a function $f_k : \mathbb{R} \to \{1, \cdots, M_k\}$, where M_k is some positive integer, and defines the number of cells in the kth side encoder. The K side encoders generate a partition $\mathcal{A}$ of $\mathbb{R}$, $\mathcal{A} = \{A_{i_1,\cdots,i_K} | (i_1, \cdots, i_K) \in \mathcal{I}_{\mathcal{K}}\}$, where $A_{i_1,\cdots,i_K} = \{x | f_1(x) = i_1, \cdots, f_K(x) = i_K\}$. The partition $\mathcal{A}$ is called the central partition.

Each index i_k is transmitted over the kth channel. Each channel has some probability of breaking down, so at the receiver, there are two kinds of situations with respect to each description: either the channel works properly and the received index is correct, or the channel breaks down and nothing is received.

Let $\mathcal{L} = \{l_1, \cdots, l_k, \cdots, l_s\} \subseteq \mathcal{K}$ denote a subset of descriptions, where $1 \leq k \leq s$, and $1 \leq s \leq K$. Assume only those descriptions in the subset $\mathcal{L}$ are received at the decoder. Then the decoder $g_{\mathcal{L}}$ corresponding to the arrived descriptions is used to reconstruct the source sample. The decoder $g_{\mathcal{L}}$ maps each s-tuple $(i_{l_1}, \cdots, i_{l_s})$ to some reconstruction level $a^{\mathcal{L}}_{i_{l_1},\cdots,i_{l_s}} \in \mathbb{R}$, where $1 \leq i_{l_k} \leq M_k$, for all $1 \leq k \leq s$.

The set of K side encoders $(f_1, \cdots, f_K)$ is denoted by $\mathbf{f}$ and is called the encoder of the MDSQ. The set of decoders $(g_{\mathcal{L}})_{\mathcal{L} \subseteq \mathcal{K}}$ is denoted by $\mathbf{g}$ and is called the decoder of the MDSQ. An MDSQ is completely determined by the encoder $\mathbf{f}$ and the decoder $\mathbf{g}$. Note that $\mathbf{f}$ is completely specified by the central partition $\mathcal{A}$ and the assignment of K-tuples to the set of cells in the central partition, which is called the index assignment (IA). In fixed-rate MDSQ, the rate of each side encoder, $r_k, 1 \leq k \leq K$, is the number of bits used to represent any index generated by the kth side encoder. Hence $r_k = \lceil \log_2 M_k \rceil$.

The expected distortion of the source reconstruction when only a subset $\mathcal{L}$ of descriptions is received at the decoder is

$$d_{\mathcal{L}} = \sum_{(i_{l_1},\cdots,i_{l_s}) \in \mathcal{I}_{\mathcal{L}}} \int_{\cap_{k=1}^{s} f_{l_k}^{-1}(i_{l_k})} \left(x - a^{\mathcal{L}}_{i_{l_1},\cdots,i_{l_s}} \right)^2 f_X(x)dx, \tag{8.3}$$

where $f_{l_k}^{-1}(i_{l_k})$ denotes the set of values which are mapped by f_{l_k} to the index i_{l_k}. Here $\mathcal{I}_{\mathcal{L}}$ denotes the set of all possible s-tuples of indices corresponding to $\mathcal{L}$.

The problem of optimal MDSQ design could be formulated as the problem of minimizing the central distortion $d_{\mathcal{K}}$ with constraints imposed on the rates and the other component distortions:

$$\text{minimize} \quad d_{\mathcal{K}} \tag{8.4}$$

$$\text{subject to} \quad d_{\mathcal{L}} \leq D_{\mathcal{L}}, \text{for all} \quad \mathcal{L} \subset \mathcal{K},$$
$$r_k = R_k, k = 1, 2, \cdots, K,$$

where $D_{\mathcal{L}}, \mathcal{L} \subseteq \mathcal{K}$, and $R_k, 1 \leq k \leq K$, are some fixed values. For fixed-rate MDSQ, the constraints on the rates can be satisfied by fixing the values of $M_1, \cdots, M_K$, such that $M_k = 2^{R_k}, 1 \leq k \leq K$.

The constraints on distortions can be eliminated by using the Lagrangian relaxation method. Precisely, the Lagrangian functional is defined as:

$$\mathcal{L}(\mathbf{f}, \mathbf{g}, \{\lambda_{\mathcal{L}}\}_{\mathcal{L}}) = \sum_{\mathcal{L} \subseteq \mathcal{K}} \lambda_{\mathcal{L}} d_{\mathcal{L}}, \tag{8.5}$$

where $\lambda_{\mathcal{L}} \geq 0, \mathcal{L} \subseteq \mathcal{K}$, are the Lagrangian multipliers, and $\lambda_{\mathcal{K}} = 1$. Then the problem is converted to the unconstrained optimization problem:

$$\text{minimize}_{\mathbf{f},\mathbf{g}} \quad \mathcal{L}(\mathbf{f}, \mathbf{g}, \{\lambda_{\mathcal{L}}\}_{\mathcal{L}}). \tag{8.6}$$

If there are Lagrangian multipliers such that the solution $(\mathbf{f}^*(\lambda_{\mathcal{L}}), \mathbf{g}^*(\lambda_{\mathcal{L}}))$ to the problem (8.6) satisfies the constraints in (8.4) with equality, then $(\mathbf{f}^*(\lambda_{\mathcal{L}}), \mathbf{g}^*(\lambda_{\mathcal{L}}))$ is a solution to (8.4). It is guaranteed that such Lagrangian multipliers exist only if $(\{R_k\}_k, \{D_{\mathcal{L}}\}_{\mathcal{L}})$ is on the lower convex hull of the operational MD rate-distortion region. Therefore, this conversion is able to solve optimally only some instances of the problem, while for the others it provides a good approximation.

Another variant of formulating the problem of optimal MDSQ design is that of minimizing the overall expected distortion at the receiver with constraints imposed on rates:

$$\text{minimize} \quad \sum_{\mathcal{L} \subseteq \mathcal{K}} p_{\mathcal{L}} d_{\mathcal{L}} \tag{8.7}$$

$$\text{subject to} \quad r_k = R_k, k = 1, 2, \cdots, K,$$

where $p_{\mathcal{L}}$ is defined as the probability that only those descriptions in subset $\mathcal{L}$ are available for decoding. If we set the Lagrangian multipliers in (8.5) to the values of $p_{\mathcal{L}}$'s in (8.7), then the problem (8.7) becomes equivalent to (8.6).

In the special case of symmetric descriptions, the rates of all descriptions are the same, i.e., $r_1 = \cdots = r_k = r$, and $p_{\mathcal{L}}$ only varies with the number of received descriptions. In other words, $p_{\mathcal{L}_i} = p_{\mathcal{L}_j} = p(s)$ if $|\mathcal{L}_i| = |\mathcal{L}_j| = s$.

The first algorithm to solve (8.6) was proposed by Vaishampayan in [44] for the fixed-rate case and in [98] for the entropy-constrained case. He considered the case of two descriptions, and his approach is a generalization of Lloyd algorithm for scalar quantizer design [88]. This algorithm begins with an initial encoder and proceeds iteratively, each iteration consisting of the following two steps: (1) fix the encoder and optimize the decoder, and (2) fix the decoder and optimize the encoder. The value of the Lagrangian decreases at each iteration, and since it is bounded below by zero, the sequence of Lagrangians is guaranteed to converge to a local minimum. In general, the globally optimal solution is not guaranteed to be achieved except for some special cases, namely, when cells are required to be contiguous and the pdf of the source is log-concave [78], [82]. In [45], [77], [81], [90], globally optimal algorithms for MDSQ design are proposed for discrete distributions, under the assumption that all cells are convex. The complexity of globally optimal MRSQ design without the cell convexity restriction was discussed in [84] and fast near-optimal algorithms were proposed in [83], both for the case of discrete sources.

8.2.2 Generalized Lloyd algorithm for optimal MDSQ design

Vaishampayan's algorithm [44] can be extended in a straightforward manner to the general K-description case, $K \geq 2$. In this section we describe this extension. Here we use the second approach (8.7) for formulating the problem of optimal MDSQ design. By substituting (8.3) in (8.7), the optimization problem becomes

$$\text{minimize}_{\mathbf{f},\mathbf{g}} \sum_{\mathcal{L}\subseteq\mathcal{K}} p_{\mathcal{L}} \sum_{(i_{l_1},\cdots,i_{l_s})\in\mathcal{I}_{\mathcal{L}}} \int_{\cap_{k=1}^{s} f_{l_k}^{-1}(i_{l_k})} \left(x - a_{i_{l_1},\cdots,i_{l_s}}^{\mathcal{L}}\right)^2 f_X(x)dx. \tag{8.8}$$

Decoder optimization. When the encoder is fixed, the decoder can be optimized by separately minimizing each term in the summation in (8.8). This turns into a simpler problem of separately minimizing each integral, leading to the solution:

$$a_{i_{l_1},\cdots,i_{l_s}}^{\mathcal{L}} = E[X|X \in \bigcap_{k=1}^{s} f_{l_k}^{-1}(i_{l_k})] = \frac{\int_{\cap_{k=1}^{s} f_{l_k}^{-1}(i_{l_k})} x f_X(x)dx}{\int_{\cap_{k=1}^{s} f_{l_k}^{-1}(i_{l_k})} f_X(x)dx}. \tag{8.9}$$

Encoder optimization. At this step, the decoder is fixed and the encoder is optimized. Here, it is useful to rewrite the expression in (8.8) into the equivalent form:

$$\sum_{(i_1,\cdots,i_K)\in\mathcal{I}_{\mathcal{K}}} \int_{A_{i_1,\cdots,i_K}} \left[\sum_{\mathcal{L}\subseteq\mathcal{K}} p_{\mathcal{L}} \left(x - a_{i_{l_1},\cdots,i_{l_s}}^{\mathcal{L}}\right)^2 f_X(x)\right]dx, \tag{8.10}$$

where $l_1,\cdots,l_s$, denote the elements of $\mathcal{L}$. The purpose of this step is to design the sets $A_{i_1,\cdots,i_K}$ for all $(i_1,\cdots,i_K) \in \mathcal{I}_{\mathcal{K}}$, such that they form a partition of $\mathbb{R}$ and (8.10) is minimized. A sufficient condition for minimizing (8.10) is

$$A_{i_1,\cdots,i_K} = \left\{x \in \mathbb{R}| \sum_{\mathcal{L}\subseteq\mathcal{K}} p_{\mathcal{L}} \left(x - a_{i_{l_1},\cdots,i_{l_s}}^{\mathcal{L}}\right)^2 \leq \sum_{\mathcal{L}\subseteq\mathcal{K}} p_{\mathcal{L}} \left(x - a_{i'_{l_1},\cdots,i'_{l_s}}^{\mathcal{L}}\right)^2, \right.$$
$$\left. \text{for all} \quad \left(i'_1,\cdots,i'_K\right) \in \mathcal{I}_{\mathcal{K}} - \{(i_1,\cdots,i_K)\}\right\}. \tag{8.11}$$

Further, by denoting

$$\alpha_{i_1,\cdots,i_K} = \sum_{\mathcal{L}\subseteq\mathcal{K}} p_{\mathcal{L}} a_{i_{l_1},\cdots,i_{l_s}}^{\mathcal{L}}$$

$$\beta_{i_1,\cdots,i_K} = \sum_{\mathcal{L}\subseteq\mathcal{K}} p_{\mathcal{L}} \left(a_{i_{l_1},\cdots,i_{l_s}}^{\mathcal{L}}\right)^2, \tag{8.12}$$

we can write (8.11) into another form which is similar to the one given in [44]:

$$A_{i_1,\cdots,i_K} = \{x \in \mathbb{R}| 2\alpha_{i_1,\cdots,i_K} x - \beta_{i_1,\cdots,i_K} \geq 2\alpha_{i'_1,\cdots,i'_K} x - \beta_{i'_1,\cdots,i'_K},$$
$$\text{for all} \quad \left(i'_1,\cdots,i'_K\right) \in \mathcal{I}_{\mathcal{K}} - \{(i_1,\cdots,i_K)\}\}. \tag{8.13}$$

It is clear that $A_{i_1,\cdots,i_K}$ is either an interval or an empty set. Let us order the elements of $\mathcal{I}_{\mathcal{K}}$ in increasing order of $\alpha_{i_1,\cdots,i_K}$. If for two distinct K-tuples $(i_1,\cdots,i_K)$ and $(i'_1,\cdots,i'_K)$, we have $\alpha_{i_1,\cdots,i_K} = \alpha_{i'_1,\cdots,i'_K}$, then either (1) $\beta_{i_1,\cdots,i_K} \neq \beta_{i'_1,\cdots,i'_K}$ or (2) $\beta_{i_1,\cdots,i_K} = \beta_{i'_1,\cdots,i'_K}$. In the first case, if $\beta_{i_1,\cdots,i_K} < \beta_{i'_1,\cdots,i'_K}$, then the set $A_{i'_1,\cdots,i'_K}$ is empty, hence the K-tuple $(i'_1, \cdots, i'_K)$ should be eliminated from $\mathcal{I}_{\mathcal{K}}$. Conversely, if $\beta_{i_1,\cdots,i_K} > \beta_{i'_1,\cdots,i'_K}$, then $(i_1, \cdots, i_K)$ should be eliminated from $\mathcal{I}_{\mathcal{K}}$. We can conclude that for the first case, one K-tuple has to be eliminated from the index set $\mathcal{I}_{\mathcal{K}}$. For the second case, we have $A_{i_1,\cdots,i_K} = A_{i'_1,\cdots,i'_K}$, hence one of the two K-tuples can also be eliminated from $\mathcal{I}_{\mathcal{K}}$. Therefore, after removing from $\mathcal{I}_{\mathcal{K}}$ the K-tuples with identical $\alpha_{i_1,\cdots,i_K}$'s, $\mathcal{I}_{\mathcal{K}}$ has the property that for any two distinct K-tuples in $\mathcal{I}_{\mathcal{K}}$, we have $\alpha_{i_1,\cdots,i_K} \neq \alpha_{i_1,\cdots,i_K}$.

Let us order the K-tuples in $\mathcal{I}_{\mathcal{K}}$ in increasing order of $\alpha_{i_1,\cdots,i_K}$. Now consider the mapping $\pi : \mathcal{I}_{\mathcal{K}} \to \{1, 2, \cdots, N\}$ which maps each K-tuple $(i_1, \cdots, i_K)$ to an integer representing its position in this ordering. Hence π is characterized by the property that $\pi((i_1, \cdots, i_K)) < \pi\left((i'_1, \cdots, i'_K)\right)$ if and only if $\alpha_{i_1,\cdots,i_K} < \alpha_{i'_1,\cdots,i'_K}$. Further, we use the notation α_ℓ, β_ℓ and A_ℓ, respectively, as short forms for $\alpha_{i_1,\cdots,i_K}$, $\beta_{i_1,\cdots,i_K}$ and $A_{i_1,\cdots,i_K}$, respectively, where $\ell = \pi((i_1, \cdots, i_K))$. Also, for $m < n$, denote

$$x_{m,n} = \frac{\beta_n - \beta_m}{2(\alpha_n - \alpha_m)}. \tag{8.14}$$

Then we obtain

$$
\begin{aligned}
A_1 &= \left(-\infty, \min_{1 < n \leq N} x_{1,n}\right], \\
A_N &= \left[\max_{1 \leq \ell < N} x_{\ell,N}, \infty\right), \quad \text{and} \\
A_i &= \left[\max_{1 \leq \ell < i} x_{\ell,i}, \min_{i < n \leq N} x_{i,n}\right], \quad \text{for} \quad 1 < i < N,
\end{aligned}
\tag{8.15}
$$

with the convention that $[b, a] = \emptyset$ if $b > a$.

The cells in the optimal central partition can be formed directly by solving (8.15) in $O(N^2)$ time. Vaishampayan uses a more ingenious method to identify the empty sets, thus being able to decrease the amount of computations for some inputs. Very recently, a more efficient algorithm which runs in $O(N)$ time has been proposed in [80].

8.2.3   Index assignment

Because Vaishampayan's algorithm can only guarantee a locally optimal solution, the choice of the initial encoder influences its performance. The problem of finding a good initial encoder is strongly related to the problem of finding a good IA.

It was shown in [44] that for an optimal MDSQ, the cells in the central partition must be convex in the case of squared error distortion measure. Therefore, the cells in the central partition are intervals, while those in the side partitions are unions of possibly disconnected intervals. The central partition can be described by a vector of thresholds $\mathbf{t} = (t_0, t_1, \cdots, t_N)$, where $t_0 = -\infty$, $t_N = \infty$ and each interval $(t_{\ell-1}, t_\ell]$ is a cell in the central partition. The IA can be formally defined as a one-to-one mapping h from the set of indices ℓ of each interval $(t_{\ell-1}, t_\ell]$, $1 \leq \ell \leq N$, to the set $\mathcal{I}_{\mathcal{K}}$ of K-tuples $(i_1, \cdots, i_K)$, hence $h : \{1, \cdots, N\} \to \mathcal{I}_{\mathcal{K}}$, where $\mathcal{I}_{\mathcal{K}} \subseteq \{1, \cdots, M_1\} \times \cdots \times \{1, \cdots, M_k\}$. Precisely, $h(\ell) = (i_1, \cdots, i_K)$ means that $(t_{\ell-1}, t_\ell] = A_{i_1,\cdots,i_K}$. The vector of thresholds $\mathbf{t}$ and the IA together fully describe the encoder.

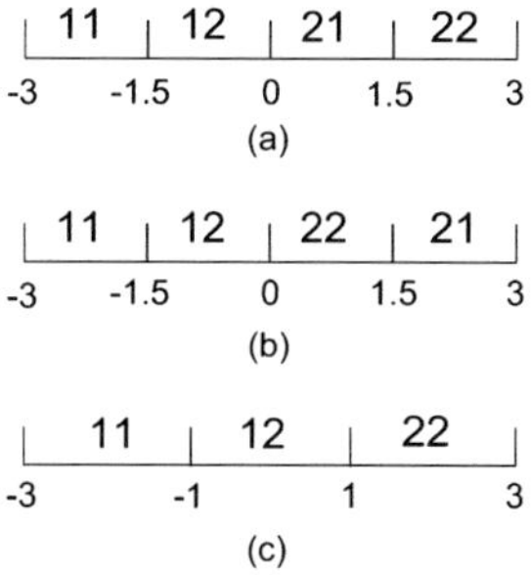

$$
\begin{array}{|c|c|c|c|}
\hline
11 & 12 & 21 & 22 \\
\hline
\end{array}
$$

| 11 | 12 | 21 | 22 |
-3   -1.5   0   1.5   3

(a)

| 11 | 12 | 22 | 21 |
-3   -1.5   0   1.5   3

(b)

| 11 | 12 | 22 |
-3   -1   1   3

(c)

Figure 8.2 Three MDSQ's for a uniformly distributed source

Table 8.1 Central and side distortions

	$E(d^{12})$	$E(d^1)$	$E(d^2)$
(a)	0.1876	0.75	2.4376
(b)	0.1876	0.75	3
(c)	0.3333	1	1

Note that when designing the IA, there are two aspects to be taken into consideration: (1) the choice of the subset $\mathcal{I}_{\mathcal{K}} \subseteq \{1, \cdots, M_1\} \times \cdots \times \{1, \cdots, M_k\}$, and (2) the mapping h. In [44], good IA's are proposed for the case of two balanced descriptions.

In order to shed some light on the importance of the IA, we give an example to show how the central and side distortions can be influenced by IA in the case of two descriptions. Assume the source samples are uniformly distributed over the interval $[-3, 3]$ and that $r_1 = r_2 = 1$ bit per source sample. Consider the three MDSQ's illustrated in Fig. 8.2 with different central partition and IA. The two-digit number in each central cell denotes this cell's index pair: the first digit is the index generated by side encoder f_1, and the second digit is the index generated by side encoder f_2. Table 8.1 gives the central and side distortions of the three quantizers in Fig. 8.2.

Both (a) and (b) have the same central partition, hence their central distortions are the same. However, case (a) is superior to (b) because it has lower side distortion, which suggests that for the same set of index pairs, different mapping may result in different distortions. Then let us compare (a) with (c). (c) has 1 less central cell than (a), hence higher central distortion, but (c) has much lower average side distortion than (a). This example suggests that the number of cells (N) in the central partition induces the tradeoff between the central and side distortions: as the number of central cells decreases, the central distortion will increase, and the average of side distortions will decrease.

The design of IA for the case of two balanced descriptions, proposed by Vaishampayan [44], includes two parts: the selection of the set of index pairs and the mapping of those pairs to the central cells. Assume the rate for each description is $r = \lceil \log_2 M \rceil$, and the number of central cells is $N(N \leq M^2)$. Let $\ell \min_p(n), p = 1, 2$, be the minimum value of the interval index ℓ that is mapped to an index pair whose pth element is equal to n. Let $\ell \max_p(n), p = 1, 2$ be the maximum value of the interval index ℓ that is mapped to an index pair whose pth element is equal to n. Then we define the spread of the nth cell of the pth side encoder

$$s_p(n) = \ell \max_p(n) - \ell \min_p(n) + 1, \quad p = 1, 2. \tag{8.16}$$

It has been proposed in [44] that to design a good IA, the spread $s_p(n)$ needs to be minimized, assuming that it is constant with respect to n and p. The set of index pairs $\mathcal{I}_{\mathcal{K}}$ can be identified with the elements of an $M \times M$ matrix (the pair (i_1, i_2) corresponds to the element on the i_1th row and i_2th column). Then, the choice of $\mathcal{I}_{\mathcal{K}}$ can be regarded as selecting a set of elements of the matrix. Let us assume that the index pairs $(1, 1), (2, 2), \cdots, (M, M)$ are always in the set, and they are assigned to central cells from left to right. This way, the index pairs $(1, 1), (2, 2), \cdots, (M, M)$ form the main diagonal of the $M \times M$ matrix. To minimize the spread, we would exhaust all elements on a diagonal which lies closer to the main diagonal before moving to a diagonal which lies further away from the main diagonal. Hence, we select a set of index pairs which lie on the main diagonal and on the $2k$ diagonals closest to the main diagonal. The mapping of the selected set of index pairs to the central cells could be interpreted as deciding a scanning sequence of this set to minimize the spread. Two families of IA's are presented in [44] for the case of two balanced descriptions: modified nested IA and modified linear IA. Figure 8.3 shows examples proposed in [44] of these two IA families with $k = 3$, $M = 8$.

For the case of more descriptions, there is not a well-developed methodology for constructing good IA's. The work [74] presents an IA for MDSQ with more descriptions, but only when the side and central distortions are of interest (no other joint quantizers' distortions). [97] introduced a new method for MDSQ design, called the sequential design of MDSQ.

1	3	5	7				
2	8	10	12	14			
4	9	15	17	19	21		
6	11	16	22	23	25	27	
	13	18	24	29	30	32	34
		20	26	31	36	37	39
			26	33	38	41	43
				35	40	42	44

(a) Modified nested IA

1	2	4	7				
3	5	8	11	14			
6	9	12	15	18			
10	13	16	19	21	24	28	
	17	20		23	27	31	35
			22	26	30	34	38
			25	29	33	37	40
				32	36	39	41

(b) Modified linear IA

Figure 8.3 IA examples for k=3, M=8 [44]. © 1993 IEEE.

8.3   Lattice MDVQ

Theoretically, MDVQ can achieve the MDC rate distortion bound as block length approaches infinity. Unfortunately, optimal MDVQ design is computationally intractable (optimal single-description VQ design is already NP-Hard [86]). A practical way of managing the complexity is to use lattice VQ codebooks. This reduces the MDVQ design problem to one of choosing a lattice Λ for central description and an associated sublattice Λ_s for $K \geq 2$ side descriptions, and establishing a one-to-one mapping, called index assignment α, between any point $\lambda \in \Lambda$ and an ordered K-tuple $(\lambda_1, \ldots, \lambda_K) \in \Lambda_s^K$. The above MDVQ scheme was first proposed by Servetto *et al.* [94], and commonly referred to as multiple-description lattice vector quantization (MDLVQ). Given the dimension of source vectors, lattices Λ and Λ_s can be selected from the known optimal and/or near-optimal lattice vector quantizers (e.g., those tabulated in [129]). Therefore, the key issue in optimal MDLVQ design is to find the bijection function $\alpha : \Lambda \leftrightarrow \alpha(\Lambda) \subset \Lambda_s^K$ that minimizes a distortion measure weighted over all possible channel/network scenarios. The seminal paper of [99] studied the index assignment problem for $K = 2$ balanced MDLVQ in considerable length, and proposed a "guiding principle" for constructing an optimal index assignment for two balanced descriptions. Also, the authors pointed out that optimal MDLVQ index assignment is a problem of linear assignment. However, a challenging algorithmic problem remains. This is how to reduce the graph matching problem from an association between two infinite sets Λ and Λ_s^K to between a finite subset of Λ and a finite subset of Λ_s^K, and keep these two finite sets as small as possible without compromising optimality.

Diggavi *et al.* proposed a technique of converting the index assignment problem for two-description lattice VQ to a finite bipartite graph matching problem [75]. Two sublattices Λ_1, Λ_2, and their product sublattice of Λ_s, are used to construct the two description LVQ. The index assignment is obtained by a minimum weight matching between a Voronoi set of central lattice points and a set of edges (ordered pairs of sublattice points, one end point in Λ_1 and the other in Λ_2). Each set has a cardinality of $N_1 N_2$, where N_k is the index of Λ_k, $k = 1, 2$. Therefore, the index assignment can be computed in $O((N_1 N_2)^{5/2})$ time, given that the weighted bipartite graph matching can be solved in $O(N^{5/2})$ time [130].

In [75] the authors only argued their index assignment algorithm to be optimal for two-description lattice scalar quantizers, and left its optimality for lattice vector quantizers unexamined. This technique of constructing MDLVQ using a product sublattice was extended from two descriptions to any K balanced descriptions by Østergaard *et al.* [91]. Østergaard *et al.* also used linear assignment to find index assignments. Their solution seemed to require $O(N^5)$ time, where N is the sublattice index, because it used a candidate set of $O(N^2)$ central lattice points. Even with such a large set of candidate central lattice points, still no bound was given on the size of the candidate K-tuples of sublattice points used for labeling, and no proof of optimality was offered.

In [132] Huang and Wu proposed an $O(N)$ greedy index assignment algorithm for MDLVQ of any $K \geq 2$ balanced descriptions in any dimensions. They proved that the algorithm minimizes the expected distortion given the loss probability p and entropy rate R_s of side descriptions, as $N \to \infty$. Moreover, they showed that for $K = 2$, under some mild conditions, the algorithm is optimal for finite N as well. For $K > 2$ and a finite N, the greedy algorithm was augmented by a fast local adjustment procedure and it was conjectured that the augmented algorithm is optimal in general.

The remainder of this chapter is structured as follows. The next section formulates the optimal MDLVQ design problem and introduces necessary notations. Section 8.3.4 presents the greedy index assignment algorithm. An asymptotical ($N \to \infty$) optimality of the proposed algorithm is proven in Section 8.3.8. Constructing the proof leads to some new and improved closed-form expressions of the expected MDLVQ distortion in N and K, which are also presented in the section. Section 8.3.11 sharpens some results of the previous section for two balanced descriptions, by proving the optimality and deriving an exact distortion formula of the proposed algorithm for finite N. The non-asymptotical results of Section 8.3.11 use a so-called S-similar sublattice. Section 8.4 shows that common lattices in signal quantization do have S-similar sublattices. In Section 8.5, we consider that the greedy index assignment may be suboptimal for finite N when $K > 2$.

8.3.1   Preliminaries

In a K-description MDLVQ, an input vector $x \in R^L$ is first quantized to its nearest lattice point $\lambda \in \Lambda$, where Λ is a fine lattice. Then the lattice point λ is mapped by a bijective labeling function α to an ordered K-tuple $(\lambda_1, \lambda_2, \cdots, \lambda_K) \in \Lambda_s^K$, where Λ_s is a coarse lattice. Let the components of α be $(\alpha_1, \alpha_2, \cdots, \alpha_K)$, i.e., $\alpha_k(\lambda) = \lambda_k, 1 \leq k \leq K$. With the function α the encoder generates K descriptions of x: $\lambda_k, 1 \leq k \leq K$, and transmits each description via an independent channel to a receiver.

If the decoder receives all K descriptions, it can reconstruct x to λ with the inverse labeling function α^{-1}. In general, due to channel losses, the decoder receives only a subset χ of the K descriptions; then it can reconstruct x to the average of the received descriptions:

$$\hat{x} = \frac{1}{|\chi|} \sum_{\lambda_i \in \chi} \lambda_i.$$

Note the optimal decoder that minimizes the mean square error should decode x to the centroid of the points $\lambda \in \Lambda$ whose corresponding components $\alpha(\lambda)$ are in χ. But decoding to the average of received descriptions is easy for design [131]. It is also asymptotically optimal for two-description case [99].

8.3.1.1   Lattice and sublattice

A lattice Λ in the L-dimensional Euclidean space is a discrete set of points

$$\Lambda \triangleq \{\lambda \in \mathbb{R}^L : \lambda = uG, u \in \mathbb{Z}^L\}, \tag{8.17}$$

i.e., the set of all possible integral linear combinations of the rows of a matrix G. The $L \times L$ matrix G of full rank is called a generator matrix for the lattice. The Voronoi cell

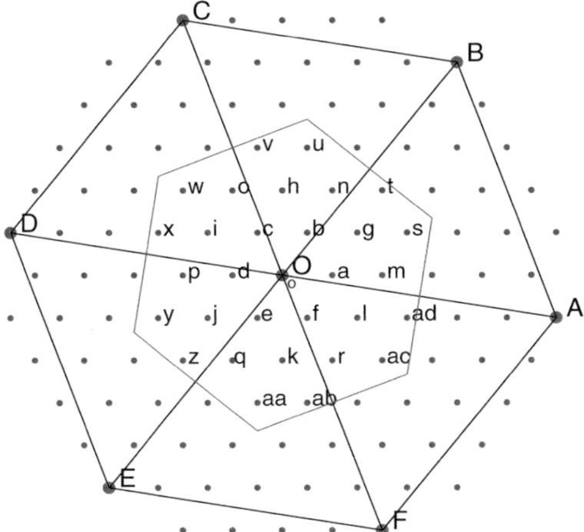

Figure 8.4 Hexagonal lattice A_2 and its sublattice with index $N = 31$. Central lattice points are marked by small dots, and sublattice points by big dots. © 2006 IEEE. Reprinted with permission from [132]

of a lattice point $\lambda \in \Lambda$ is defined as

$$V(\lambda) \triangleq \{x \in \mathbb{R}^L : \|x - \lambda\| \leqslant \|x - \tilde{\lambda}\|, \forall \tilde{\lambda} \in \Lambda\}, \tag{8.18}$$

where $\|x\|^2 = \langle x, x \rangle$ is the dimension-normalized norm of vector x.

Two lattices are used in the MDLVQ system: a fine lattice Λ and a coarse lattice Λ_s. The fine lattice Λ is the codebook for the central decoder when all the descriptions are received, thus called central lattice. The coarse lattice Λ_s is the codebook for a side decoder when only one description is received. Typically, $\Lambda_s \subset \Lambda$, hence Λ_s is also called a sublattice. The ratio of the point densities of Λ and Λ_s, which is also the ratio of the volumes of the Voronoi cells of Λ_s and Λ, is defined as the sublattice index N. If the sublattice is clean (no central lattice points lie on the boundary of a sublattice Voronoi cell), N is equal to the number of central lattice points inside a sublattice Voronoi cell. Sublattice index N governs trade offs between the side and central distortions. We assume that Λ_s is geometrically similar to Λ, i.e., Λ_s can be obtained by scaling, rotating, and possibly reflecting Λ [129]. Figure 8.4 is an example of a hexagonal lattice and its sublattice with index $N = 31$.

Let G and G_s be generator matrices for L-dimensional central lattice Λ and sublattice Λ_s. Then Λ_s is geometrically similar to Λ if and only if there exist an invertible $L \times L$ matrix U with integer entries, a scalar β, and an orthogonal $L \times L$ matrix A with determinant 1 such that

$$G_s = UG = \beta GA. \tag{8.19}$$

The index for a geometrically similar lattice is $N = \frac{det G_s}{det G} = \beta^L$.

8.3.1.2   Rate of MDLVQ

In MDLVQ, a source vector x of joint pdf $g(x)$ is quantized to its nearest fine lattice $\lambda \in \Lambda$. The probability of quantizing x to a lattice λ is

$$P(\lambda) = \int_{V(\lambda)} g(x)dx. \qquad (8.20)$$

The entropy rate per dimension of the output of the central quantizer is [99]

$$
\begin{aligned}
R_c &= \frac{1}{L} \sum_{\lambda \in \Lambda} P(\lambda) \log P(\lambda) \\
&= -\frac{1}{L} \sum_{\lambda \in \Lambda} \int_{V(\lambda)} g(x)dx \log_2 \int_{V(\lambda)} g(x)dx \\
&\approx -\frac{1}{L} \sum_{\lambda \in \Lambda} \int_{V(\lambda)} g(x)dx \log_2 g(\lambda)v \\
&= h(p) - \frac{1}{L} \log_2 v,
\end{aligned}
\qquad (8.21)
$$

where v is the volume of a Voronoi cell of Λ, and $h(p)$ is the differential entropy. The above assumes high resolution when $g(x)$ is approximately constant within a Voronoi cell $V(\lambda)$.

The volume of a Voronoi cell of the sublattice Λ_s is $v_s = Nv$. Denote by $Q(x) = \lambda$ the quantization mapping. Then, similar to (8.21), the entropy rate per dimension of a side description (for balanced MDLVQ) is [99]

$$
\begin{aligned}
R_s &= \frac{1}{L}H(\alpha_k(Q(X))) \\
&\approx h(p) - \frac{1}{L} \log_2 v_s \\
&= h(p) - \frac{1}{L} \log_2(Nv).
\end{aligned}
\qquad (8.22)
$$

The total entropy rate per dimension for the balanced MDLVQ system is

$$R_t = KR_s. \qquad (8.23)$$

8.3.2   Distortion of MDLVQ

Assuming that the K channels are independent and each has a failure probability p, we can write the expected distortion as

$$D = \sum_{k=0}^{K} \binom{K}{k}(1-p)^k p^{K-k} D_k,$$

where D_k is the expected distortion when receiving k out of K descriptions.

For the case of all descriptions received, the average distortion per dimension is given by

$$d_c = \sum_{\lambda \in \Lambda} \int_{V(\lambda)} \|x - \lambda\|^2 g(x)dx \approx G_\Lambda v^{\frac{2}{L}}, \qquad (8.24)$$

where G_Λ is the dimensionless normalized second moment of lattice Λ [129]. The approximation is under the standard high-resolution assumption.

If only description i is received, the expected side distortion is [99]

$$
\begin{aligned}
d_i &= \sum_{\lambda \in \Lambda} \int_{V(\lambda)} \|x - \lambda_i\|^2 \, g(x) dx \\
&= \sum_{\lambda \in \Lambda} \int_{V(\lambda)} \left(\|x - \lambda\|^2 + \|\lambda - \lambda_i\|^2 + 2\langle x - \lambda, \lambda - \lambda_i \rangle \right) g(x) dx \\
&\approx d_c + \sum_{\lambda \in \Lambda} \|\lambda - \lambda_i\|^2 P(\lambda), \quad 1 \le i \le K.
\end{aligned}
\tag{8.25}
$$

Hence, the expected distortion when receiving only one description is

$$
D_1 = \frac{1}{K} \sum_{i=1}^{K} d_i = d_c + \sum_{\lambda \in \Lambda} \frac{1}{K} \sum_{i=1}^{K} \|\lambda - \lambda_i\|^2 P(\lambda).
\tag{8.26}
$$

Let m_K be the centroid of all K descriptions $\lambda_1, \lambda_2, \cdots, \lambda_K$, that is,

$$
m_K \triangleq \frac{1}{K} \sum_{k=1}^{K} \lambda_k.
\tag{8.27}
$$

Then we have

$$
\begin{aligned}
\frac{1}{K} \sum_{i=1}^{K} \|\lambda - \lambda_i\|^2 &= \frac{1}{K} \sum_{i=1}^{K} \|(\lambda - m_K) - (\lambda_i - m_K)\|^2 \\
&= \|\lambda - m_K\|^2 + \frac{1}{K} \sum_{i=1}^{K} \|\lambda_i - m_K\|^2 - \frac{2}{K} \left\langle \lambda - m_K, \sum_{i=1}^{K} (\lambda_i - m_K) \right\rangle \\
&= \|\lambda - m_K\|^2 + \frac{1}{K} \sum_{i=1}^{K} \|\lambda_i - m_K\|^2.
\end{aligned}
\tag{8.28}
$$

Substituting (8.28) into (8.26), we get

$$
D_1 = d_c + \sum_{\lambda \in \Lambda} \left(\|\lambda - m_K\|^2 + \frac{1}{K} \sum_{i=1}^{K} \|\lambda_i - m_K\|^2 \right) P(\lambda).
\tag{8.29}
$$

Now we consider the case of receiving k descriptions, $1 < k < K$. Let I be the set of all possible combinations of receiving k out of K descriptions. Let $\iota = (\iota_1, \iota_2, \cdots, \iota_k)$ be an element of I. Under high-resolution assumption, we have [91]

$$
\begin{aligned}
D_k &= d_c + |I|^{-1} \sum_{\lambda \in \Lambda} \sum_{\iota \in I} \left\| \lambda - \frac{1}{k} \sum_{j=1}^{k} \lambda_{\iota_j} \right\|^2 P(\lambda) \\
&= d_c + \sum_{\lambda \in \Lambda} \left(\|\lambda - m_K\|^2 + \frac{K - k}{(K-1)k} \frac{1}{K} \sum_{i=1}^{K} \|\lambda_i - m_K\|^2 \right) P(\lambda), \quad 1 < k < K.
\end{aligned}
\tag{8.30}
$$

Substituting the expressions of D_k into (8.24), we arrive at

$$D = (1 - p^K)d_c + \sum_{\lambda \in \Lambda} \left(\zeta_1 \|\lambda - m_K\|^2 + \zeta_2 \frac{1}{K} \sum_{i=1}^{K} \|\lambda_i - m_K\|^2 \right) P(\lambda) + p^K E[\|X\|^2],$$

(8.31)

where

$$\zeta_1 = \sum_{k=1}^{K-1} \binom{K}{k} (1 - p)^k p^{K-k} = 1 - p^K - (1 - p)^K$$

(8.32)

$$\zeta_2 = \sum_{k=1}^{K-1} \binom{K}{k} (1 - p)^k p^{K-k} \frac{K - k}{(K - 1)k}.$$

8.3.3 Optimal MDLVQ design

Given source and channel statistics and given total entropy rate R_t, optimal MDLVQ design involves (i) the choice of the central lattice Λ and the sublattice Λ_s; (ii) the determination of optimal number of descriptions K and of the optimal sublattice index value N; and (iii) the optimization of index assignment function α once (i) and (ii) are fixed. We defer the discussions of optimal values of K and N to Section 8.3.8, and first focus on the construction of optimal index assignment. It turns out that our new constructive approach will lead to improved analytical results of K and N in optimal MDLVQ design.

With fixed p, K, Λ, Λ_s, the optimal MDLVQ design problem (i.e., minimizing (8.31)) reduces to finding the optimal index assignment α that minimizes the average side distortion

$$d_s \triangleq \sum_{\lambda \in \Lambda} \left(\frac{1}{K} \sum_{i=1}^{K} \|\alpha_i(\lambda) - \mu(\alpha(\lambda))\|^2 + \zeta \|\lambda - \mu(\alpha(\lambda))\|^2 \right) P(\lambda), \qquad (8.33)$$

where

$$\mu(\alpha(\lambda)) = K^{-1} \sum_{i=1}^{K} \alpha_i(\lambda)$$

$$\zeta = \frac{\zeta_1}{\zeta_2} = \frac{\sum_{k=1}^{K-1} \binom{K}{k}(1 - p)^k p^{K-k}}{\sum_{k=1}^{K-1} \binom{K}{k}(1 - p)^k p^{K-k} \frac{K-k}{(K-1)k}}. \qquad (8.34)$$

When $K = 2$, the objective function can be simplified to

$$d_s = \sum_{\lambda \in \Lambda} \left(\frac{1}{4} \|\lambda_1 - \lambda_2\|^2 + \left\| \lambda - \frac{\lambda_1 + \lambda_2}{2} \right\|^2 \right) P(\lambda). \qquad (8.35)$$

8.3.4 Index assignment algorithm

This section presents a new greedy index assignment algorithm for MDLVQ of $K \geq 2$ balanced descriptions and examines its optimality. The algorithm is very simple and it hinges on an interesting new notion of K-fraction sublattice. We first define this K-fraction sublattice and reveal its useful properties for optimizing index assignment. Then we describe the greedy index assignment algorithm, analyze its complexity, and prove its optimality for $K = 2$. The last part of this section addresses a so-called S-similarity property of sublattices that is used in the optimality proof of the greedy algorithm.

8.3.5 *K-fraction sublattice*

In the following study of optimal index assignment for K balanced descriptions, the sublattice

$$\Lambda_{s/K} \triangleq \frac{1}{K}\Lambda_s = \left\{\tau \in \mathbb{R}^L : \tau = \frac{u}{K}G_s, u \in \mathbb{Z}^L\right\} \tag{8.36}$$

plays an important role, and it will be referred to as the K-fraction sublattice hereafter.

The K-fraction sublattice $\Lambda_{s/K}$ has the following interesting relations to Λ and Λ_s.

PROPERTY 8.1 $\mu(\alpha(\lambda)) = K^{-1}\sum_{k=1}^{K}\alpha_k(\lambda)$ is an onto (but not one-to-one) map: $\Lambda_s^K \to \Lambda_{s/K}$.

Proof   (1) $(\lambda_1, \lambda_2, \cdots, \lambda_K) \in \Lambda_s^K \Rightarrow \sum_{k=1}^{K}\lambda_k \in \Lambda_s \to K^{-1}\sum_{k=1}^{K}\lambda_k \in \Lambda_{s/K}$; (2) $\forall \tau \in \Lambda_{s/K}$, let $\lambda_1 = K\tau, \lambda_2 = \cdots = \lambda_K = 0$, then $\lambda_1, \lambda_2, \cdots, \lambda_K \in \Lambda_s$ and $\mu(\alpha(\lambda)) = \tau$. $\square$

This means that the centroid of any K-tuples in Λ_s^K must be in $\Lambda_{s/K}$, and further $\Lambda_{s/K}$ consists only of these centroids.

If two K-fraction sublattice points τ_1, τ_2 satisfy $\tau_1 - \tau_2 \in \Lambda_s$, then we say that τ_1 and τ_2 are in the same coset with respect to Λ_s. Any K-fraction sublattice point belongs to one of the cosets.

PROPERTY 8.2 $\Lambda_{s/K}$ has, in the L-dimensional space, K^L cosets with respect to Λ_s.

Proof   Let τ_1, τ_2 be two K-fraction sublattice points. τ_1, τ_2 can be expressed by

$$\tau_1 = \frac{u}{K}G_s, \ \tau_2 = \frac{v}{K}G_s,$$

where $u = (u_1, u_2, \cdots, u_L) \in \mathbb{Z}^L, v = (v_1, v_2, \cdots, v_L) \in \mathbb{Z}^L$. Two points τ_1 and τ_2 fall in the same coset with respect to Λ_s if and only if $u_i \equiv v_i \bmod K$ for all $i = 1, 2, \cdots, L$. The claim follows since the remainder of division by K takes on K different values. $\square$

The K-fraction sublattice $\Lambda_{s/K}$ partitions the space into Voronoi cells. Denote the Voronoi cell of a point $\tau \in \Lambda_{s/K}$ by

$$V_{s/K}(\tau) = \{x : \|x - \tau\| \le \|x - \tilde{\tau}\|, \forall \tilde{\tau} \in \Lambda_{s/K}\}.$$

PROPERTY 8.3 $\Lambda_{s/K}$ is clean, if Λ_s is clean.

Proof Assume for a contradiction that there was a point $\lambda \in \Lambda$ on the boundary of $V_{s/K}(\tau)$ for a $\tau \in \Lambda_{s/K}$. Scaling both λ and $V_{s/K}(\tau)$ by K places $K\lambda$ on the boundary of $KV_{s/K}(\tau) = \{Kx : \|Kx - K\tau\| \leq \|Kx - K\tilde{\tau}\|, \forall \tilde{\tau} \in \Lambda_{s/K}\}$. But $K\lambda$ is a point of Λ, and $KV_{s/K}(\tau)$ is nothing but the Voronoi cell V_s of the sublattice point $K\tau \in \Lambda_s$, or the point $K\lambda \in \Lambda$ lies on the boundary of $V_s(K\tau)$, contradicting that Λ_s is clean. $\square$

PROPERTY 8.4   Both lattices Λ_s and Λ are symmetric about any point $\tau \in \Lambda_{s/2}$.

Proof $\forall \tau \in \Lambda_{s/2}$, we have $2\tau \in \Lambda_s$, so $2\tau - \lambda_s \in \Lambda_s$ holds for $\forall \lambda_s \in \Lambda_s$; similarly, $\forall \tau \in \Lambda_{s/2}$, we have $2\tau \in \Lambda$, so $2\tau - \lambda \in \Lambda$ holds for $\forall \lambda \in \Lambda$. $\square$

8.3.6   Greedy index assignment algorithm

Our motive for constructing the K-fraction lattice $\Lambda_{s/K}$ is to relate $\Lambda_{s/K}$ to the central lattice Λ in such a way that the two terms of d_s in (8.33) can be minimized independently. This is brought to light by examining the partition of the space by Voronoi cells of K-fraction sublattice points. For simplicity, we assume the sublattice Λ_s is clean (if not, the algorithm still works by employing a rule to break a tie on the boundary of a sublattice Voronoi cell). According to Property 8.3, no point $\lambda \in \Lambda$ is on the boundary of any Voronoi cell of $\Lambda_{s/K}$. Let

$$\text{B}(\tau) = \left\{(\lambda_1, \lambda_2, \cdots, \lambda_K) \in \Lambda_s^K | \Sigma_{1 \leq k \leq K} \lambda_k / K = \tau \right\} \tag{8.37}$$

be the set of all ordered K-tuples of sublattice points of centroid τ, and $\tau \in \Lambda_{s/K}$ by Property 8.1.

In constructing an index assignment, we sort the members of $\text{B}(\tau)$ by $\sum_{k=1}^{K} \|\lambda_k - \tau\|^2$. From $\text{B}(\tau)$ we select the ordered K-tuples in increasing values of $\sum_{k=1}^{K} (\lambda_k - \tau)^2$ to label the central lattice points inside the K-fraction Voronoi cell $V_{s/K}(\tau)$, until all $N_\tau = |\Lambda \cap V_{s/K}(\tau)|$ of those central lattice points are labeled. It follows from (8.33) that any bijective mapping between the $n(\tau)$ central lattice points and the N_τ ordered K-tuples of sublattice points yields the same value of d_s. Such an index assignment clearly minimizes the second term of (8.33), which is the sum of the squared distances of all central lattice points in Voronoi cell $V_{s/K}(\tau)$ to the centroid $\tau = \mu(\alpha(\cdot))$. As $N \rightarrow \infty$, the proposed index assignment algorithm also minimizes the first term of (8.33) independently. This will be proven with some additional effort in Section 8.3.8.

For the two-description case, these N_τ ordered pairs are formed by the N_τ nearest sublattice points to τ in Λ_s by Property 8.4. Note when $\tau \in \Lambda_s$, the ordered pair (τ, τ) should be used to label τ itself.

According to Property 8.2, $\Lambda_{s/K}$ has K^L cosets with respect to Λ_s in the L-dimensional space, so there are K^L classes of $V_{s/K}(\tau)$. We only need to label one representative out of each class, and cover the whole space by shifting. Thus it suffices to label a total of N central lattice points.

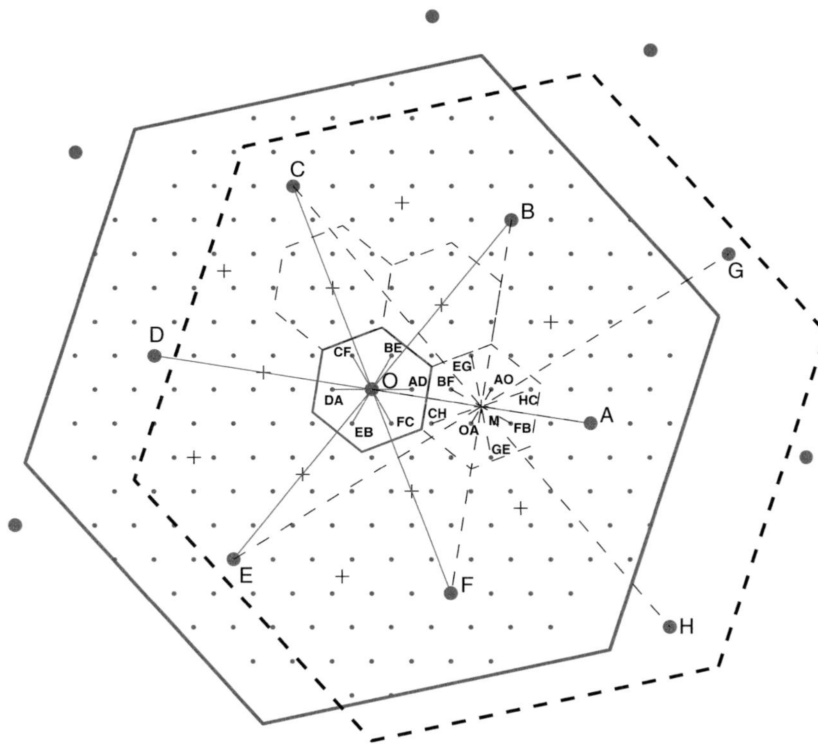

Figure 8.5 Optimal index assignments for A_2 lattice with $N = 31$, $K = 2$. Points of Λ, Λ_s and $\Lambda_{s/2}$ are marked by $\cdot$, $\bullet$, and $+$, respectively. © 2006 IEEE. Reprinted with permission from [132]

8.3.7 Examples of greedy index assignment algorithm

To visualize the work of the proposed index assignment algorithm, let us examine two examples on an A_2 lattice (see Figs. 8.5 and 8.6). The A_2 lattice Λ is generated by basis vectors represented by complex numbers: 1 and $\omega = 1/2 + i\sqrt{3}/2$. By shifting invariance of A_2 lattice, we only need to label the N central lattice points that belong to K^2 Voronoi cells of $\Lambda_{s/K}$. By angular symmetry of A_2 lattice, we can further reduce the number of points to be labeled.

The first example is a two-description case, with the sublattice Λ_s given by basis vectors $5 - \omega$, $\omega(5 - \omega)$, which is geometrically similar to Λ, has index $N = 31$, and is clean (refer to Fig. 8.5). There are two types of Voronoi cells of $\Lambda_{s/2}$, as shown by the solid and dashed boundaries in Fig. 8.5. The solid cell is centered at a central lattice point and contains 7 central lattice points. The dashed cell is centered at the midpoint of the line segment OA, and contains 8 central lattice points. To label the 7 central lattice points in $V_{s/2}(O)$, we use the 7 nearest sublattice points to O: (O, A, B, C, D, E, F). They form 6 ordered pairs with the midpoint O: $(A, D), (D, A), (B, E), (E, B), (C, F), (F, C)$, and an unordered pair (O, O) since O is itself a sublattice point. To label the 8 central

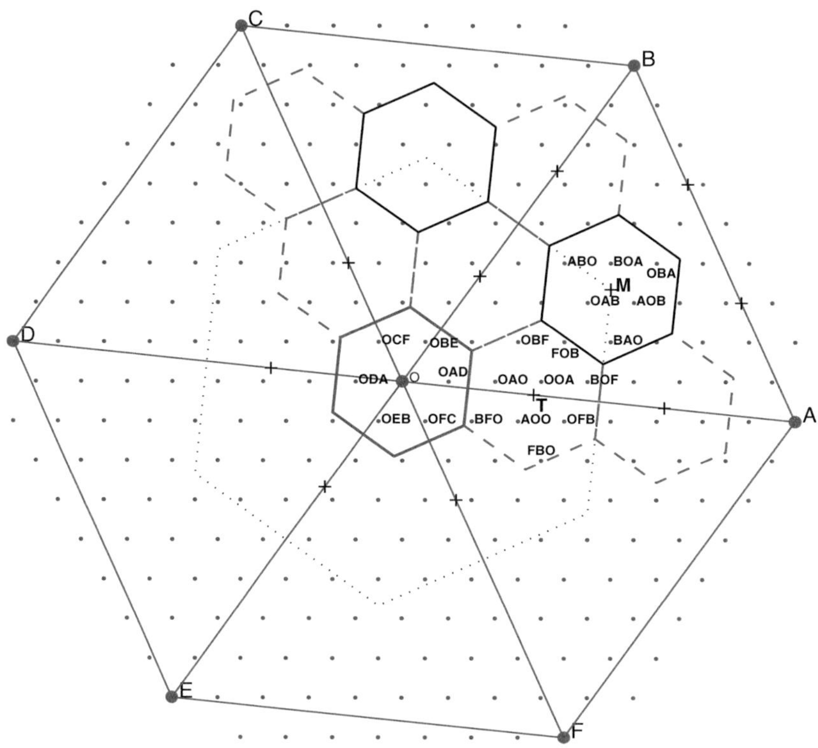

Figure 8.6 Optimal index assignments for A_2 lattice with $N = 73, K = 3$. Points of Λ, Λ_s, and $\Lambda_{s/3}$ are marked by $\cdot$, $\bullet$, and $+$, respectively. © 2006 IEEE. Reprinted with permission from [132]

lattice points in $V_{s/2}(M)$, we use the 8 nearest sublattice points to M: (O, A, B, F, C, H, E, G). They form 8 ordered pairs with midpoint M: (O, A), (A, O), (B, F), (F, B), (C, H), (H, C), (E, G), (G, E). The labeling of the 7 central lattice points in $V_{s/2}(O)$ and the labeling of the 8 central lattice points in $V_{s/2}(M)$ are illustrated in Fig. 8.5.

Figure 8.6 illustrates the result of the proposed algorithm in the case of three descriptions. The depicted index assignment for three balanced descriptions is computed for the sublattice of index $N = 73$ and basis vectors: $8 - \omega$, $\omega(8 - \omega)$.

The presented MDLVQ index assignment algorithm is fast with an $O(N)$ time complexity. The simplicity and low complexity of the algorithm are due to the greedy optimization approach adopted.

The tantalizing question is, of course, can the greedy algorithm be optimal? A quick test on the above two examples may be helpful. Let the distance between a nearest pair of central lattice points in Λ be one. For the first example, the result of [99] (the best so far) is $d_s = 561/31 = 18.0968$, while the greedy algorithm does better, producing $d_s = 528/31 = 17.0323$. Indeed, in both examples, one can verify that the expected distortion is minimized as the two terms of d_s in (8.33) are minimized independently.

8.3.8 Asymptotically optimal design of MDLVQ

In this section we first prove that the greedy index assignment is optimal for any K, p, Λ and Λ_s as $N \to \infty$. In constructing the proof, we derive a close-form asymptotical expression of the expected distortion of optimal MDLVQ for general $K \geq 2$. It allows us to determine the optimal volume of a central lattice Voronoi cell ν, the optimal sublattice index N, and the optimal number of descriptions K, given the total entropy rate of all side descriptions R_t and the loss probability p. These results, in addition to optimal index assignment α, complete the design of optimal MDLVQ, and they present an improvement over previous work of [91].

8.3.9 Asymptotical optimality of the proposed index assignment

Since the second term of d_s is minimized by the Voronoi partition defined by the K-fraction lattice, the optimality of the proposed index assignment based on the K-fraction lattice follows if it also minimizes the first term of d_s. This is indeed the case when $N \to \infty$. To compute the first term of d_s, let

$$\varsigma_k \triangleq \sum_{i=1}^{k} \lambda_i, \quad k = 1, 2, \cdots, K.$$

Then

$$\sum_{k=1}^{K} \|\lambda_k - \tau\|^2 = \sum_{k=1}^{K} \left\| \lambda_k - \frac{1}{K}\varsigma_K \right\|^2$$

$$= \left(\sum_{k=1}^{K-1} \left\| \lambda_k - \frac{1}{K}\varsigma_K \right\|^2 \right) + \left\| \lambda_K - \frac{1}{K}\varsigma_K \right\|^2$$

$$= \left(\sum_{k=1}^{K-1} \left\| \left(\lambda_k - \frac{1}{K-1}\varsigma_{K-1} \right) + \frac{1}{K-1}\left(\varsigma_{K-1} - \frac{K-1}{K}\varsigma_K \right) \right\|^2 \right)$$

$$\quad + \left\| \varsigma_{K-1} - \frac{K-1}{K}\varsigma_K \right\|^2$$

$$\overset{(a)}{=} \left(\sum_{k=1}^{K-1} \left\| \lambda_k - \frac{1}{K-1}\varsigma_{K-1} \right\|^2 \right) + \frac{K}{K-1}\left\| \varsigma_{K-1} - \frac{K-1}{K}\varsigma_K \right\|^2$$

$$\overset{(b)}{=} \sum_{k=1}^{K-1} \frac{k+1}{k}\left\| \varsigma_k - \frac{k}{k+1}\varsigma_{k+1} \right\|^2. \tag{8.38}$$

Equality (a) holds because the inner product $\left\langle \sum_{k=1}^{K-1}\left(\lambda_k - \frac{1}{K-1}\varsigma_{K-1} \right), \frac{1}{K-1}\left(\varsigma_{K-1} - \frac{K-1}{K}\varsigma_K \right) \right\rangle$ is zero. After using the same deduction $K-1$ times, we arrive at equality (b).

Note the one-to-one correspondence between $(\lambda_1, \lambda_2, \cdots, \lambda_K)$ and $(\varsigma_1, \varsigma_2, \cdots, \varsigma_K)$. Also recall that the proposed index assignment uses the N_τ (the number of central lattice points in K-fraction Voronoi cell $V_{s/K}(\tau)$) smallest K-tuples in $\mathcal{B}(\tau)$ according to the value of $\sum_{k=1}^{K} \|\lambda_k - \tau\|^2$. Finding the N_τ smallest values of $\sum_{k=1}^{K} \|\lambda_k - \tau\|^2$ in $\mathcal{B}(\tau)$ is equivalent to finding the N_τ smallest values of $\sum_{k=1}^{K-1} \frac{k+1}{k} \left\| \varsigma_k - \frac{k}{k+1} \varsigma_{k+1} \right\|^2$ among the $(K-1)$-tuples $(\varsigma_1, \varsigma_2, \cdots, \varsigma_{K-1})$ with $\varsigma_K = \sum_{k=1}^{K} \lambda_k$.

THEOREM 8.1 *The proposed greedy index assignment algorithm is optimal as $N \to \infty$ for any given Λ, Λ_s, K, and p.*

Proof The ith nearest sublattice point to $\frac{k}{k+1} \varsigma_{k+1}$ is approximately on the boundary of an L-dimensional sphere with volume $iN\nu$. Given ς_{k+1}, the ith smallest value of $\left\| \varsigma_k - \frac{k}{k+1} \varsigma_{k+1} \right\|^2$ is approximately $(iN\nu/B_L)^{\frac{2}{L}}/L = G_L \left(1 + \frac{2}{L}\right)(iN\nu)^{\frac{2}{L}}$, where $B_L = G_L^{-\frac{L}{2}}(L+2)^{-\frac{L}{2}}$ is the volume of an L-dimensional sphere of unit radius [99], and G_L is the dimensionless normalized second moment of an L-dimensional sphere.

Let $f^{(n)}(\tau)$ be the nth smallest value of $\sum_{k=1}^{K} \|\lambda_k - \tau\|^2$ in $\mathcal{B}(\tau)$ that is realized at $\left(\varsigma_1^{(n)}, \varsigma_2^{(n)}, \cdots, \varsigma_{K-1}^{(n)}\right)$. Then

$$f^{(n)}(\tau) = \sum_{k=1}^{K-1} \frac{k+1}{k} \left\| \varsigma_k^{(n)} - \frac{k}{k+1} \varsigma_{k+1}^{(n)} \right\|^2$$

$$\approx G_L \left(1 + \frac{2}{L}\right)(N\nu)^{\frac{2}{L}} \sum_{k=1}^{K-1} \frac{k+1}{k} \left(i_k^{(n)}\right)^{\frac{2}{L}}, \qquad (8.39)$$

in which $\left(i_1^{(n)}, i_2^{(n)}, \cdots, i_{K-1}^{(n)}\right) \in \mathbb{Z}^{K-1}$ is where the sum $\sum_{k=1}^{K-1} \frac{k+1}{k}(i_k)^{\frac{2}{L}}$ takes on its nth smallest value over all $(K-1)$-tuples of positive integers.

When $N \to \infty$, the proposed index assignment algorithm takes the $N_\tau \approx N/K^L$ smallest terms of $\sum_{k=1}^{K} \|\lambda_k - \tau\|^2$ in $\mathcal{B}(\tau)$ for every τ. But (8.39) states that the nth smallest value of $\sum_{k=1}^{K} \|\lambda_k - \tau\|^2$ is independent of τ. Therefore, the first term of d_s is minimized, establishing the optimality of the resulting index assignment. $\qquad \square$

Remark 8.1 The $O(N)$ MDLVQ index assignment algorithm based on the K-fraction lattice is so far the only one proven to be asymptotically optimal, except for the prohibitively expensive linear assignment algorithm. In the next section, we will strengthen the above proof in a constructive perspective, and establish the optimality of the algorithm for finite N when $K = 2$.

8.3.10   Optimal design parameters ν, N, and K

Now our attention turns to the determination of the optimal ν (the volume of a Voronoi cell of Λ), N (the sublattice index), and K (the number of descriptions) that achieve minimum expected distortion, given the total entropy rate of all side descriptions R_t and loss probability p.

Using (8.39), we have

$$\sum_{\lambda\in\Lambda}\sum_{k=1}^{K}\|\lambda_k-\mu(\alpha(\lambda))\|^2 P(\lambda) = \sum_{\tau\in\Lambda_{s/K}}\sum_{\lambda\in V_{s/K}(\tau)}\sum_{k=1}^{K}\|\lambda_k-\mu(\alpha(\lambda))\|^2 P(\lambda)$$

$$\approx \frac{1}{N_\tau}\sum_{n=1}^{N_\tau}\sum_{k=1}^{K-1}\frac{k+1}{k}\left\|\varsigma_k^{(n)}-\frac{k}{k+1}\varsigma_{k+1}^{(n)}\right\|^2$$

$$\approx G_L\left(1+\frac{2}{L}\right)(Nv)^{\frac{2}{L}}\frac{1}{N_\tau}\sum_{n=1}^{N_\tau}\sum_{k=1}^{K-1}\frac{k+1}{k}\left(i_k^{(n)}\right)^{\frac{2}{L}}.$$

$$(8.40)$$

Consider the region defined as

$$\Omega \triangleq \left\{\sum_{k=1}^{K-1}\frac{k+1}{k}x_k^{\frac{2}{L}}\leq C\,|x_1,x_2,\cdots,x_{K-1}\geq 0, x_1,x_2,\cdots,x_{K-1}\in\mathbb{R}\right\}. \quad (8.41)$$

Choose C appropriately so that the volume of Ω is $V(\Omega)=N_\tau$. As $N_\tau\to\infty$, Ω contains approximately N_τ optimal integer vectors $(i_1, i_2,\cdots, i_{K-1})$. These N_τ points are uniformly distributed in Ω, with density one point per unit volume. Because the ratio between the volume occupied by each point and the total volume is $1/N_\tau$, which approaches zero when $N_\tau\to\infty$, we can replace the summation by integral and get

$$\frac{1}{N_\tau}\sum_{n=1}^{N_\tau}\sum_{k=1}^{K-1}\frac{k+1}{k}\left(i_k^{(n)}\right)^{\frac{2}{L}} \approx \frac{\int_{x\in\Omega}\sum_{k=1}^{K-1}\frac{k+1}{k}x_k^{\frac{2}{L}}\,dx}{\int_{x\in\Omega}dx}$$

$$= \frac{\int_{y\in\Omega_0}\sum_{k=1}^{K-1}y_k^{\frac{2}{L}}\,dy}{\int_{y\in\Omega_0}dy}, \quad (8.42)$$

where $y_k = \left(\frac{k+1}{k}\right)^{\frac{L}{2}}x_k, k=1,2,\cdots,K-1$, and Ω_0 is defined as

$$\Omega_0 \triangleq \left\{\sum_{k=1}^{K-1}y_k^{\frac{2}{L}}\leq C\,|y_1,y_2,\cdots,y_{K-1}\geq 0, y_1,y_2\cdots,y_{K-1}\in\mathbb{R}\right\}. \quad (8.43)$$

Substituting (8.42) into (8.40), we have

$$\sum_{\lambda\in\Lambda}\sum_{k=1}^{K}\|\lambda_k-\mu(\alpha(\lambda))\|^2 P(\lambda) \approx G_L\left(1+\frac{2}{L}\right)(Nv)^{\frac{2}{L}}\frac{\int_{y\in\Omega_0}\sum_{k=1}^{K-1}y_k^{\frac{2}{L}}\,dy}{\int_{y\in\Omega_0}dy}. \quad (8.44)$$

Let $V(\Omega_0)$ be the volume of region Ω_0, i.e.,

$$
\begin{aligned}
V(\Omega_0) &= \int_{y \in \Omega_0} dy_1 \, dy_2 \cdots dy_{K-1} \\
&= K^{\frac{L}{2}} \int_{x \in \Omega} dx_1 \, dx_2 \cdots dx_{K-1} \\
&= K^{\frac{L}{2}} N_\tau \\
&= K^{-\frac{L}{2}} N,
\end{aligned}
\tag{8.45}
$$

and define the dimensionless normalized $\frac{2}{L}$th moment of Ω_0:

$$
G_{\Omega_0} \triangleq \frac{1}{K-1} \frac{\int_{y \in \Omega_0} \sum_{k=1}^{K-1} y_k^{\frac{2}{L}} \, dy}{V(\Omega_0)^{1 + \frac{2}{L(K-1)}}}.
\tag{8.46}
$$

Note that scaling Ω_0 does not change G_{Ω_0}. For the special case $L = 1$, the region Ω_0 is a $(K-1)$-dimensional sphere in the first octant, so the normalized second moment $G_{\Omega_0} = 4G_{K-1}$. For the special case $K = 2$, $G_{\Omega_0} = L/(L+2)$ is the normalized $\frac{2}{L}$th moment of a line $[0, C]$. Generally, using *Dirichlet's Integral* [46], we get

$$
G_{\Omega_0} = \frac{1}{n + \frac{2}{L}} \frac{\Gamma\left(\frac{nL}{2} + 1\right)^{\frac{2}{nL}}}{\Gamma\left(\frac{L}{2} + 1\right)^{\frac{2}{L}}}.
\tag{8.47}
$$

Hence,

$$
\frac{\int_{y \in \Omega_0} \sum_{k=1}^{K-1} y_k^{\frac{2}{L}} \, dy}{\int_{y \in \Omega_0} dy} = G_{\Omega_0}(K-1) V(\Omega_0)^{\frac{2}{L(K-1)}} = G_{\Omega_0}(K-1) K^{\frac{-1}{K-1}} N^{\frac{2}{L(K-1)}}.
\tag{8.48}
$$

Substituting (8.48) into (8.44), we have

$$
\frac{1}{K} \sum_{\lambda \in \Lambda} \sum_{k=1}^{K} \|\lambda_k - \mu(\alpha(\lambda))\|^2 P(\lambda) \approx G_L G_{\Omega_0} \left(1 + \frac{2}{L}\right)(K-1) K^{\frac{-K}{K-1}} N^{\frac{2K}{L(K-1)}} v^{\frac{2}{L}}
$$

$$
\approx G_L \Phi_{K-1,L}(K-1) K^{\frac{-K}{K-1}} N^{\frac{2K}{L(K-1)}} v^{\frac{2}{L}},
\tag{8.49}
$$

where

$$
\Phi_{n,L} = \frac{1 + \frac{2}{L}}{n + \frac{2}{L}} \frac{\Gamma\left(\frac{nL}{2} + 1\right)^{\frac{2}{nL}}}{\Gamma\left(\frac{L}{2} + 1\right)^{\frac{2}{L}}}.
$$

Note $\Phi_{n,1} = 12G_n$ and $\Phi_{1,L} = 1$.

When $N \to \infty$, $N_\tau \approx N/K^L$ independently of the cell center τ. The N_τ central lattice points are uniformly distributed in $V_{s/K}(\tau)$, whose volume is approximately $N_\tau v$. Hence the second term of d_s can be evaluated as

$$\zeta \sum_{\lambda \in \Lambda} \|\lambda - \mu(\alpha(\lambda))\|^2 P(\lambda) \approx \zeta G_\Lambda (N_\tau \nu)^{\frac{2}{L}}$$

$$= \zeta G_\Lambda K^{-2} (N\nu)^{\frac{2}{L}}. \tag{8.50}$$

Comparing (8.50) with (8.49), the first term of d_s dominates the second term when $N \to \infty$, thus

$$d_s \approx G_L \Phi_{K-1,L}(K-1)K^{\frac{-K}{K-1}} N^{\frac{2K}{L(K-1)}} \nu^{\frac{2}{L}}. \tag{8.51}$$

Substituting (8.51) and (8.24) into (8.31), we finally express the expected distortion of optimal MDLVQ in a closed form:

$$D \approx (1 - p^K)G_\Lambda \nu^{\frac{2}{L}} + \zeta_2 G_L \Phi_{K-1,L}(K-1)K^{\frac{-K}{K-1}} N^{\frac{2K}{L(K-1)}} \nu^{\frac{2}{L}} + p^K E[\|X\|^2]. \tag{8.52}$$

Using a different index assignment algorithm Østergaard *et al.* derived a similar expression for the expected MDLVQ distortion (equation (51) in [91]):

$$D^* \approx (1 - p^K)G_\Lambda \nu^{\frac{2}{L}} + \hat{K}G_L \psi_L^2 N^{\frac{2K}{L(K-1)}} \nu^{\frac{2}{L}} + p^K E[\|X\|^2], \tag{8.53}$$

where

$$\hat{K} = \sum_{k=1}^{K-1} \binom{K}{k}(1-p)^k p^{K-k} \frac{K-k}{2kK} \tag{8.54}$$

and ψ_L is a quantity that is given analytically only for $K = 2$ and for $K = 3$ with odd L and is determined empirically for other cases.

To compare D and D^*, we rewrite (8.52) as

$$D \approx (1 - p^K)G_\Lambda \nu^{\frac{2}{L}} + \hat{K}G_L \hat{\psi}_L^2 N^{\frac{2K}{L(K-1)}} \nu^{\frac{2}{L}} + p^K E[\|X\|^2], \tag{8.55}$$

where

$$\hat{\psi}_L = \sqrt{2K^{\frac{-1}{K-1}} \Phi_{K-1,L}}. \tag{8.56}$$

The two expressions are the same when $K = 2$ for which $\hat{\psi}_L = \psi_L = 1$, but they differ for $K > 2$. Table 8.2 lists the values of ψ_L and $\hat{\psi}_L$ for $K = 3$, and it shows that $\hat{\psi}_\infty = \psi_\infty = \left(\frac{4}{3}\right)^{\frac{1}{4}}$, and $\hat{\psi}_L < \psi_L$ for other values of L. This implies $D < D^*$, or that our index assignment makes the asymptotical expression of D tighter.

Now we proceed to derive the optimal value of N, which governs the optimal tradeoff between the central and side distortions for given p and K. For the total target entropy rate $R_t = KR$, we rewrite (8.21) to get

$$N\nu = 2^{L(h(p) - R_t/K)}. \tag{8.57}$$

For simplicity, define

$$\eta \triangleq 2^{L(h(p) - R_t/K)}, \tag{8.58}$$

Table 8.2 Values of ψ_L and $\hat{\psi}_L$ in L for $K = 3$. Values of ψ_L are reproduced from Table 1 in [91]. © 2006 IEEE

L	ψ_L	$\hat{\psi}_L \ldots$
1	$1.1547\ldots$	$0.9549\ldots$
2	$1.1481\ldots$	$0.9428\ldots$
3	$1.1346\ldots$	$0.9394\ldots$
5	$1.1241\ldots$	$0.9400\ldots$
7	$1.1173\ldots$	$0.9431\ldots$
9	$1.1125\ldots$	$0.9466\ldots$
11	$1.1089\ldots$	$0.9498\ldots$
13	$1.1060\ldots$	$0.9527\ldots$
15	$1.1036\ldots$	$0.9552\ldots$
17	$1.1017\ldots$	$0.9575\ldots$
19	$1.1000\ldots$	$0.9596\ldots$
21	$1.0986\ldots$	$0.9614\ldots$
51	$1.0884\ldots$	$0.9763\ldots$
71	$1.0856\ldots$	$0.9807\ldots$
101	$1.0832\ldots$	$0.9848\ldots$
∞	$1.0746\ldots$	$1.0746\ldots$

and we have

$$D = (1 - p^K)G_\Lambda \nu^{\frac{2}{L}} + \zeta_2 G_L \Phi_{K-1,L}(K-1)K^{\frac{-K}{K-1}} \eta^{\frac{2K}{L(K-1)}} \nu^{\frac{-2}{L(K-1)}} + p^K E[\|X\|^2]. \quad (8.59)$$

Differentiating D with respect to ν yields the optimal ν value:

$$\nu_{opt} = \eta \left(\frac{\zeta_2}{1 - p^K} \frac{G_L}{G_\Lambda} \frac{\Phi_{K-1,L}}{K^{\frac{K}{K-1}}} \right)^{\frac{L(K-1)}{2K}}.$$

Substituting ν_{opt} into (8.57), we get optimal N:

$$N_{opt} = \left(\frac{1 - p^K}{\zeta_2} \frac{G_\Lambda}{G_L} \frac{K^{\frac{K}{K-1}}}{\Phi_{K-1,L}} \right)^{\frac{L(K-1)}{2K}}. \quad (8.60)$$

If $K = 2$, the expression of N_{opt} can be simplified as

$$N_{opt} = \left(\frac{2(1+p)}{p} \frac{G_\Lambda}{G_L} \right)^{\frac{L}{4}}. \quad (8.61)$$

Remark 8.2 N_{opt} is independent of the total target entropy rate R_t and source entropy rate $h(p)$. It only depends on the loss probability p and on the number of descriptions K. Substituting ν_{opt} into (8.59), the average distortion can be expressed as a function of K. Then optimal K can be solved numerically.

Remark 8.3   When $K = 2$, (8.51) can be simplified to

$$d_s \approx \frac{1}{4} G_L (N^2 \nu)^{\frac{2}{L}}. \quad (8.62)$$

For any $a \in (0, 1)$, let $N = 2^{L(aR+1)}$, then $v = 2^{L(h(p)-(a+1)R-1)}$. Since $R \to \infty$ implies $N \to \infty$, substituting the expressions of N and v into (8.24) and (8.62), we get

$$\lim_{R \to \infty} d_c 2^{2R(1+a)} = \frac{1}{4} G_\Lambda 2^{2h(p)}$$

$$\lim_{R \to \infty} d_k 2^{2R(1-a)} = G_L 2^{2h(p)}, \quad k = 1, 2.$$

Therefore, the proposed MDLVQ algorithm asymptotically achieves the second-moment gain of a lattice for the central distortion, and the second-moment gain of a sphere for the side distortion, which is the same as the expression in [99]. In other words, our algorithm realizes the MDC performance bound for two balanced descriptions.

8.3.11 Non-asymptotical optimality for $K = 2$

In this section we sharpen the results of the previous section, by proving non-asymptotical (i.e., with respect to a finite N) optimality and deriving an exact distortion formula of our MDLVQ design algorithm for $K = 2$ balanced descriptions, under mild conditions. The following analysis is constructive and hence more useful than an asymptotical counterpart because the value of N is not very large in practice [87].

8.3.12 A non-asymptotical proof

Our non-asymptotical proof is built upon the following definitions and lemmas.

DEFINITION 8.1 A sublattice Λ_s is said to be centric, if the sublattice Voronoi cell $V_s(\lambda)$ centered at $\lambda \in \Lambda_s$ contains the N nearest central lattice points to λ.

Figures 8.5 and 8.6 show two examples of centric sublattices.

To prove the optimality of the greedy algorithm, we need some additional properties.

LEMMA 8.1 *Assume the sublattice Λ_s is centric. If $\lambda \in V_{s/2}(\tau)$ and $\tilde{\lambda} \notin V_{s/2}(\tilde{\tau})$, where $\lambda, \tilde{\lambda} \in \Lambda$ and $\tau, \tilde{\tau} \in \Lambda_{s/2}$, then $\|\lambda - \tau\| \leq \|\tilde{\lambda} - \tilde{\tau}\|$.*

Proof Scaling both λ and $V_{s/2}(\tau)$ by 2 places the lattice point 2λ in $V_s(2\tau)$; scaling both $\tilde{\lambda}$ and $V_{s/2}(\tilde{\tau})$ by 2 places the lattice point $2\tilde{\lambda} \notin V_s(2\tilde{\tau})$. Since a sublattice Voronoi cell contains the nearest central lattice points, $\|2\lambda - 2\tau\| \leq \|2\tilde{\lambda} - 2\tilde{\tau}\|$, and hence $\|\lambda - \tau\| \leq \|\tilde{\lambda} - \tilde{\tau}\|$. $\qquad\square$

DEFINITION 8.2 A sublattice Λ_s is said to be S-similar to Λ, if Λ_s can be generated by scaling and rotating Λ around any point $\tau \in \Lambda_{s/2}$ and $\Lambda_s \subset \Lambda$.

Note that the S-similarity requires that the center of symmetry be a point in $\Lambda_{s/2}$.

In what follows we assume that sublattice Λ_s is S-similar to Λ. Also, we denote by V_τ the region created by scaling and rotating $V_{s/2}(\tau)$ around τ.

LEMMA 8.2 *If $\lambda_s \in V_\tau$ and $\tilde{\lambda}_s \notin V_{\tilde{\tau}}$, where $\lambda_s, \tilde{\lambda}_s \in \Lambda_s$ and $\tau, \tilde{\tau} \in \Lambda_{s/2}$, then $\|\lambda_s - \tau\| \leq \|\tilde{\lambda}_s - \tilde{\tau}\|$.*

Proof This lemma follows from Lemma 8.1 and the definition of S-similar. $\qquad\square$

LEMMA 8.3 $\forall \tau \in \Lambda_{s/2}$, *the sublattice points in* V_τ *form* $|\Lambda \cap V_{s/2}(\tau)|$ *nearest ordered 2-tuples with their midpoints being* τ.

Proof   Letting $\tilde{\tau} = \tau$ in Lemma 8.2, we see that V_τ contains the $|\Lambda_s \cap V_\tau| = |\Lambda \cap V_{s/2}(\tau)|$ nearest sublattice points to τ. And these sublattice points are symmetric about τ according to Property 8.4. Thus this lemma holds.   □

THEOREM 8.2   *The proposed index assignment algorithm is optimal for* $K = 2$ *and any* N, *if the sublattice is centric and S-similar to the associated central lattice.*

Proof   By Property 8.1, for any $\lambda_1, \lambda_2 \in \Lambda_s$, $(\lambda_1 + \lambda_2)/2 \in \Lambda_{s/2}$. Now referring to (8.35), the proposed algorithm minimizes the second term $\sum_{\lambda \in \Lambda} \|\lambda - (\lambda_1 + \lambda_2)/2\|^2 P(\lambda)$ of d_s, since it labels any central lattice point $\lambda \in V_{s/2}(\tau)$ by $(\lambda_1, \lambda_2) \in \Lambda_s^2$, and $(\lambda_1 + \lambda_2)/2 = \tau$.

The algorithm also independently minimizes the first term $\sum_{\lambda \in \Lambda} \frac{1}{4}\|\lambda_1 - \lambda_2\|^2 P(\lambda)$ of d_s. Assume that $\sum_{\lambda \in \Lambda} \|\lambda_1 - \lambda_2\|^2 P(\lambda)$ was not minimized. Then there exists an ordered 2-tuple $(\tilde{\lambda}_1, \tilde{\lambda}_2) \in \Lambda_s^2$ which is not used in the index assignment, and $\|\tilde{\lambda}_1 - \tilde{\lambda}_2\| < \|\lambda_1 - \lambda_2\|$, where $(\lambda_1, \lambda_2) \in \Lambda_s^2$ is used in the index assignment. Let $\tau = (\lambda_1 + \lambda_2)/2, \tilde{\tau} = (\tilde{\lambda}_1 + \tilde{\lambda}_2)/2$. Since (λ_1, λ_2) is used to label a central lattice point in $V_{s/2}(\tau)$, $\lambda_1, \lambda_2 \in V_\tau$ by Lemma 8.3. However, $\tilde{\lambda}_1, \tilde{\lambda}_2 \notin V_{\tilde{\tau}}$, otherwise $(\tilde{\lambda}_1, \tilde{\lambda}_2)$ would be used in the index assignment by Lemma 8.3. So we have $\|\lambda_1 - \tau\| \leq \|\tilde{\lambda}_1 - \tilde{\tau}\|$ by Lemma 8.2, hence $\|\lambda_1 - \lambda_2\| \leq \|\tilde{\lambda}_1 - \tilde{\lambda}_2\|$, contradicting $\|\tilde{\lambda}_1 - \tilde{\lambda}_2\| < \|\lambda_1 - \lambda_2\|$.   □

Remark 8.4   A sublattice Voronoi cell being centric is not a necessary condition for the optimality of the greedy algorithm. For instance, for the A_2 lattice generated by basis vectors 1 and $\omega = 1/2 + i\sqrt{3}/2$ and the sublattice of index $N = 91$ that is generated by basis vectors $9 - \omega$, $\omega(9 - \omega)$, a sublattice Voronoi cell does not contain the N nearest central lattices, but the greedy algorithm is still optimal as the two terms of d_s are still independently minimized. This is shown in Fig. 8.7.

Remark 8.5   It is easy to choose a centric sublattice for relatively small N and in high dimensional lattices. For instance, the sublattices of A_2 lattice shown in Figs. 8.5 and 8.6 as well as any sublattice of Z lattice are centric.

8.3.13   Exact distortion formula for $K = 2$

We have derived an asymptotical expected distortion formula (8.52) of the proposed MDLVQ design, which improved a similar result in [91]. But so far no exact non-asymptotical expression of the expected MDLVQ distortion is known even for balanced two descriptions. This subsection presents progress on this account.

LEMMA 8.4   *If the sublattice is clean and S-similar, then the second term of* d_s *for the proposed optimal MDLVQ design for* $K = 2$ *is*

$$\sum_{\lambda \in \Lambda} \|\lambda - m_{1,2}\|^2 P(\lambda) = \frac{1}{4L} \frac{\sum_{i=1}^{N} a_i}{N}, \tag{8.63}$$

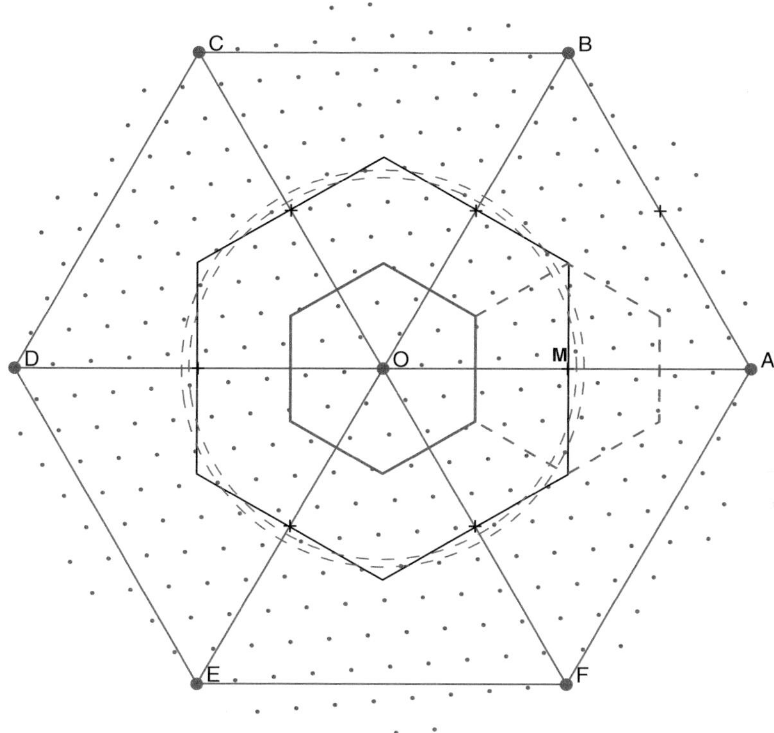

Figure 8.7 The greedy algorithm is optimal for $N = 91$, although the sublattice is not centric. The 19 central lattice points in $V_{s/2}(O)$ are labeled by the 19 nearest ordered 2-tuples with centroid O. The 24 central lattice points in $V_{s/2}(M)$ are labeled by the 24 nearest ordered 2-tuples with centroid M. Let the edge length of 2-tuple (O, A) be one: $||O - A|| \triangleq 1$. The 19th (20th) nearest ordered 2-tuple with centroid O has edge length 4 ($2\sqrt{7}$). The 24th (25th) nearest ordered 2-tuple with centroid M has edge length 5 ($3\sqrt{3}$). Because $4 < 3\sqrt{3}$ and $5 < 2\sqrt{7}$, the first term of d_s is minimized. The second term of d_s is minimized because the greedy algorithm partitions the space by the Voronoi cells of the K-fraction sublattice. © 2006 IEEE. Reprinted with permission from [132]

where a_i is the squared distance of the ith nearest central lattice point in $V_s(0)$ to the origin.

Proof $\Lambda_{s/2}$ has 2^L cosets with respect to Λ_s in the L-dimensional space. Let $\tau_1, \tau_2, \cdots, \tau_{2^L}$ be representatives of each coset. For example, when $L = 2$, $\tau_1 = (0, 0)G_s$, $\tau_2 = \left(0, \frac{1}{2}\right) G_s$, $\tau_3 = \left(\frac{1}{2}, 0\right) G_s$, $\tau_4 = \left(\frac{1}{2}, \frac{1}{2}\right) G_s$. Denote by $V_\lambda(\tau) \triangleq V_{s/2}(\tau) \bigcap \Lambda$ the set of central lattice points in the Voronoi cell of a 2-fraction sublattice point τ. We first prove that

$$2 \left(V_\lambda(\tau_i) - \tau_i\right) \underset{i \neq j}{\cap} 2 \left(V_\lambda(\tau_j) - \tau_j\right) = \varnothing. \tag{8.64}$$

$$\bigcup_{i=1}^{2^L} 2 \left(V_\lambda(\tau_i) - \tau_i\right) = V_s(0) \cap \Lambda. \tag{8.65}$$

Here for convenience, we denote by $2V$ the set of lattice points that is generated by scaling the lattice points in Voronoi cell V by 2.

Assume that (8.64) does not hold. Then there exist $\lambda_i \in V_\lambda(\tau_i)$, $\lambda_j \in V_\lambda(\tau_j)$ such that $\lambda_i - \tau_i = \lambda_j - \tau_j$. Let $\tau_0 = \tau_i - \tau_j$, then $\tau_0 \in \Lambda_{s/2}$. We also have $\tau_0 = \lambda_i - \lambda_j$, so $\tau_0 \in \Lambda$. The sublattice Λ_s is S-similar to Λ, so properly rotating and scaling Λ around the 2-fraction sublattice point τ_0 can generate Λ_s. Rotating and scaling the central lattice point $\tau_0 \in \Lambda$ around τ_0 itself generates τ_0, so $\tau_0 = \tau_j - \tau_j \in \Lambda_s$. This contradicts that τ_i and τ_j are in different cosets with respect to Λ_s, establishing (8.64).

To prove (8.65), we first show that for any $\tau \in \Lambda_{s/2}$,

$$2(V_\lambda(\tau) - \tau) = 2\left(V_{s/2}(\tau) \cap \Lambda\right) - 2\tau = V_s(2\tau) \cap (2\Lambda) - 2\tau \overset{(a)}{\subseteq} V_s(0) \cap \Lambda. \quad (8.66)$$

Step (a) holds because $2\tau \in \Lambda$ and $V_s(2\tau) - 2\tau = V_s(0)$. Therefore,

$$\overset{2^L}{\underset{i=1}{\cup}} 2\left(V_\lambda(\tau_i) - \tau_i\right) \subseteq V_s(0) \cap \Lambda. \quad (8.67)$$

According to Property 8.3, no central lattice points lie on the boundary of a K-fraction Voronoi cell when the sublattice is clean, so the set $\cup_{i=1}^{2^L} V_\lambda(\tau_i)$ contains N different central lattice points. By (8.64), the set $\cup_{i=1}^{2^L} 2\left(V_\lambda(\tau_i) - \tau_i\right)$ has N different elements. Because the set $V_s(0) \cap \Lambda$ also has N different elements and $\cup_{i=1}^{2^L} 2\left(V_\lambda(\tau_i) - \tau_i\right) \subseteq V_s(0) \cap \Lambda$, (8.65) holds.

Finally, it follows from (8.64) and (8.65) that

$$\sum_{\lambda \in \Lambda} \|\lambda - \mu(\alpha(\lambda))\|^2 P(\lambda) = \frac{1}{4} \sum_{\lambda \in \Lambda} \|2\lambda - 2\mu(\alpha(\lambda))\|^2 P(\lambda)$$

$$\overset{(a)}{=} \frac{1}{4N} \sum_{i=1}^{4} \sum_{\lambda \in V_\lambda(\tau_i)} \|2\lambda - 2\tau_i\|^2$$

$$= \frac{1}{4N} \sum_{\lambda \in V_s(0) \cap \Lambda} \|\lambda\|^2$$

$$= \frac{1}{4L} \frac{\sum_{i=1}^{N} a_i}{N}. \quad (8.68)$$

Equality (a) holds because under high-resolution assumption, $P(\lambda)$ is the same for each central lattice point $\lambda \in \cup_{i=1}^{2^L} V_\lambda(\tau_i)$. $\square$

THEOREM 8.3   *If the sublattice is clean, S-similar and centric, then the expected distortion D of optimal two-description MDLVQ is*

$$D = (1 - p^2)G_\Lambda \nu^{\frac{2}{L}} + \frac{1}{2}p(1 - p)L^{-1}\left(1 + N^{\frac{2}{L}}\right)N^{-1}\sum_{i=1}^{N} a_i + p^2 E[\|X\|^2]. \quad (8.69)$$

Proof   By Theorem 8.1, under the stated conditions, the proposed MDLVQ design is optimal. Further, the corresponding index assignment makes the first term of d_s exactly

$N^{\frac{2}{L}}$ times the second term of d_s. Then it follows from Lemma 8.4 that the first term of d_s is

$$\sum_{\lambda \in \Lambda} \frac{1}{4} \|\lambda_1 - \lambda_2\|^2 P(\lambda) = \frac{N^{\frac{2}{L}}}{4L} \frac{\sum_{i=1}^{N} a_i}{N}. \tag{8.70}$$

Substituting (8.24), (8.63), and (8.70) into (8.31), we obtain the formula of the expected distortion D in (8.69). $\qquad\square$

The above equations lead to some interesting observations. When the sublattice is centric, a_i is also the squared distance of the ith nearest central lattice point to the origin. The term $N^{\frac{2}{L}} N^{-1} \sum_{i=1}^{N} a_i$ is the average squared distance of the N nearest sublattice points to the origin, which was also realized by previous authors [99]. The other term $N^{-1} \sum_{i=1}^{N} a_i$ is the average squared distance of central lattice points in $V_s(0)$ to the origin.

The optimal v and N for a given entropy rate of side descriptions can be found by using (8.69) and $R_s = h(p) - \frac{1}{L} \log_2 Nv$ (shown in (8.22)), rather than solving many instances of index assignment problem for varying N.

8.4　*S*-similarity

The above non-asymptotical optimality proof requires the S-similarity of the sublattice. In this section we show that many commonly used lattices for signal quantization, such as A_2, Z, Z^2, $Z^L(L = 4k)$, and Z^L (L odd), have S-similar sublattices.

Being geometrically similar is a necessary condition of being S-similar, but being clean is not (for example a geometrically similar sublattice of A_2 with index 21 is S-similar, but not clean). The geometrical similar and clean sublattices of A_2, Z, Z^2, $Z^L(L = 4k)$, and Z^L (L odd) lattices are discussed in [75]. We will discuss the S-similar sublattices of these lattices in this section.

THEOREM 8.4　*For the Z lattice Λ, a sublattice Λ_s is S-similar to Λ, if and only if its index N is odd.*

Proof　Staightforward and omitted. $\qquad\square$

THEOREM 8.5　*For the A_2 lattice Λ, a sublattice Λ_s is S-similar to Λ, if it is geometrically similar to Λ and clean.*

Proof　Let Λ_s be a sublattice geometrically similar to Λ and clean. We refer to the hexagonal boundary of a Voronoi cell in Λ (respectively in Λ_s) as Λ-gon (respectively Λ_s-gon). Any point $\tau \in \Lambda_{s/2}$ is either in Λ_s or the midpoint of a Λ_s-gon edge. For instance, in Fig. 8.5 M is both the midpoint of a Λ-gon edge and the midpoint of a Λ_s-gon edge.

If $\tau \in \Lambda_s$, then $\tau \in \Lambda$, hence scaling and rotating Λ around τ yields Λ_s in this case. If τ is the midpoint of a Λ_s-gon edge, then $\tau \notin \Lambda$ because sublattice Λ_s is clean, but $\tau \in \Lambda_{1/2}$, so τ is the midpoint of a Λ-gon edge, hence scaling and rotating Λ around τ yields Λ_s in this case. $\qquad\square$

The $Z^L (L = 4l, l \geq 1)$ lattice has a geometrically similar and clean sublattice with index N, if and only if $N = m^{\frac{L}{2}}$, where m is odd [75]. Here we show that there are S-similar sublattices for at least half of these N values.

THEOREM 8.6   *The $Z^L (L = 4l, l \geq 1)$ lattice Λ has an S-similar, clean sublattice with index N, if $N = m^{\frac{L}{2}}$ with $m \equiv 1 \bmod 4$.*

Proof   We begin with the case $L = 4$. By Lagrange's four-square theorem, there exist four integers a, b, c, d such that $m = a^2 + b^2 + c^2 + d^2$. The matrix G_ξ constructed by *Lipschitz integral quaternions* $\{\xi = a + bi + cj + dk\}$ [75] is

$$
G_\xi = \begin{pmatrix} a & b & c & d \\ -b & a & d & -c \\ -c & -d & a & b \\ -d & c & -b & a \end{pmatrix}.
$$

The lattice Λ_s generated by matrix $G_s = G_\xi$ is a geometrically similar sublattice of Λ.

Let $\lambda = u$, $\lambda_s = u_s G_\xi$, $\tau = \frac{1}{2} u_\tau G_\xi$ be a point of $\Lambda, \Lambda_s, \Lambda_{s/2}$ respectively, where $u, u_s, u_\tau \in \mathbb{Z}^L$. Then,

$$
\lambda_s - \tau = \left(u_s - \frac{1}{2} u_\tau \right) G_\xi.
$$

Let $\tilde{u} = u - u_\tau \frac{1}{2} (G_\xi - I_L)$, where I_L is an $L \times L$ identity matrix, then

$$
\lambda - \tau = \tilde{u} - \frac{1}{2} u_\tau.
$$

Since $n^2 \equiv 1 \bmod 4$ or $n^2 \equiv 0 \bmod 4$ depending on whether n is an odd or even integer, $m \equiv 1 \bmod 4$ implies that exactly one of a, b, c, d is odd. Letting a be odd and b, c, d even, then $\frac{1}{2}(G_\xi - I_L)$ is an integer matrix. Hence $\tilde{u} \in \mathbb{Z}^L$. Thus, scaling and rotating Λ around point τ by scaling factor $\beta = m^{1/2}$ and rotation matrix $A = m^{-1/2} G_\xi$ yields Λ_s, proving Λ_s is S-similar to Λ.

For the dimension $L = 4l, l > 1$, let the $4l \times 4l$ generator matrix of the sublattice Λ_s be

$$
G_s = \begin{pmatrix} G_\xi & 0 & \cdots & 0 \\ 0 & G_\xi & \cdots & \vdots \\ \vdots & \vdots & \ddots & 0 \\ 0 & \cdots & 0 & G_\xi \end{pmatrix}.
$$

Then Λ_s is S-similar to Λ. And according to [75], Λ_s is clean. $\square$

The Z_2 lattice Λ has a geometrically similar sublattice Λ_s of index N, if and only if $N = a^2 + b^2, a, b \in \mathbb{Z}$. And a generator matrix for Λ_s is

$$
G_s = \begin{pmatrix} a & b \\ -b & a \end{pmatrix}. \tag{8.71}
$$

Further, Λ_s is clean if and only if N is odd [75].

THEOREM 8.7 *For the Z_2 lattice Λ, a sublattice Λ_s is S-similar to Λ, if it is geometrically similar to Λ and clean.*

Proof For a geometrically similar and clean sublattice Λ_s, its generator matrix G_s is given by (8.71). As $N = a^2 + b^2$ is odd, a and b are one even and the other odd. Letting a be odd and b even, by the same argument in proving Theorem 8.6, scaling and rotating Λ around any point $\tau \in \Lambda_{s/2}$ by scaling factor $\beta = N^{1/2}$ and rotation matrix $A = N^{-1/2}G_s$ yields Λ_s. If a is even, b is odd, scaling and rotating Λ around any point $\tau \in \Lambda_{s/2}$ by scaling factor β and rotation matrix $\tilde{A}$ yields Λ_s, where $\tilde{A}$ is an orthogonal matrix:

$$\tilde{A} = A \begin{pmatrix} 0 & -1 \\ 1 & 0 \end{pmatrix} = N^{-1/2} \begin{pmatrix} b & -a \\ a & b \end{pmatrix}.$$

$\square$

THEOREM 8.8 *An L-dimensional lattice Λ has an S-similar sublattice with index N, if $N = m^L$ is odd.*

Proof Constructing a sublattice Λ_s with index $N = m^L$ needs only scaling, i.e., $G_s = mG$. Let $\lambda = uG$, $\lambda_s = mu_sG$, $\tau = \frac{1}{2}mu_\tau G$ be in $\Lambda, \Lambda_s, \Lambda_{s/2}$ respectively, where $u, u_s, u_\tau \in \mathbb{Z}^L$. Then,

$$\lambda_s - \tau = m\left(u_s - \frac{1}{2}u_\tau\right)G.$$

Let $\tilde{u} = u - \frac{m-1}{2}u_\tau$, then $\tilde{u} \in \mathbb{Z}^L$, and

$$\lambda - \tau = \left(\tilde{u} - \frac{1}{2}u_\tau\right)G.$$

Thus, scaling Λ around point τ by $\beta = m^{1/L}$ yields Λ_s, proving Λ_s is S-similar to Λ. $\square$

COROLLARY *The Z^L (L is odd) lattice Λ has an S-similar, clean sublattice with index N, if and only if $N = m^L$ is odd.*

Proof By [75], Λ has a geometrically similar, clean sublattice of index N, if and only if $N = m^L$ is odd. A sublattice Λ_s of this index can be obtained by scaling Λ by m. Theorem 8.8 implies that Λ_s is S-similar to Λ. $\square$

8.5   Local adjustment algorithm

Theorem 8.1 is concerned with when the two terms of d_s in (8.33) can be minimized independently by the greedy index assignment algorithm. While being mostly true for $K = 2$ as stated by the theorem and as we saw in the two examples of Section 8.3.7, this may not be guaranteed when $K > 2$. Figure 8.8 presents the index assignment generated by the greedy algorithm for $K = 3$ on A_2 lattice. The solution is now suboptimal. Indeed, consider the central lattice point in $V_{s/3}(T)$ that is labeled by OAC in Fig. 8.8, changing the label from OAC to BOA will reduce d_s of the central lattice point in question. The change reduces the first term of d_s, although the second term of d_s increases

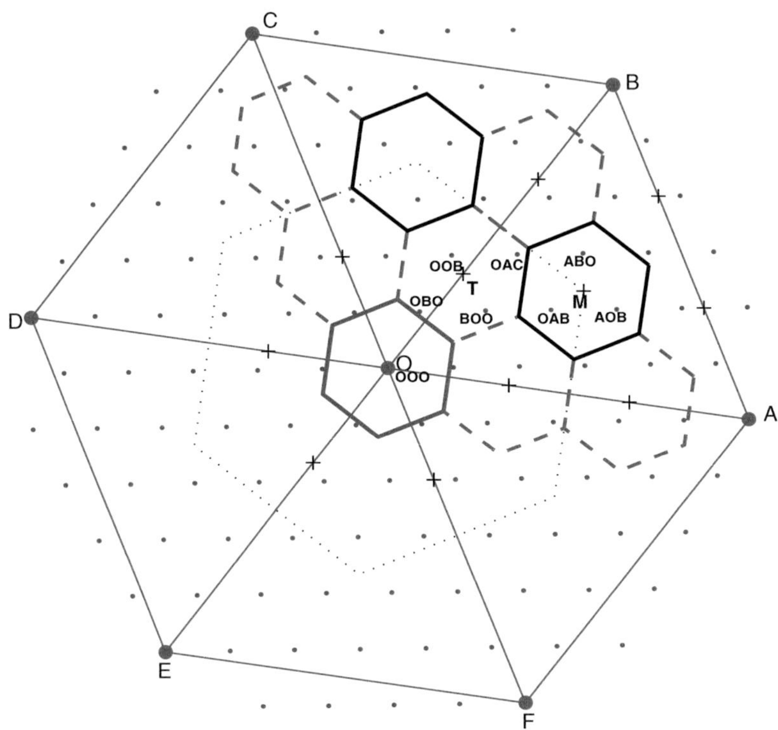

Figure 8.8 Index assignments (not optimal) by the greedy index assignment algorithm for the A_2 lattice with index $N = 31$, $K = 3$. Points of Λ, Λ_s, and $\Lambda_{s/3}$ are marked by $\cdot$, $\bullet$, and $+$, respectively. © 2006 IEEE. Reprinted with permission from [132]

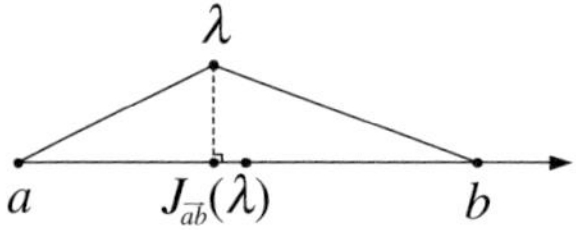

Figure 8.9 Remove lattice λ from site a, and add it to site b

slightly. Note that the 3-tuple (O, A, C) has centroid T, and the 3-tuple (B, O, A) has centroid M.

In order to make up for the loss of optimality by the greedy algorithm, we develop a local adjustment algorithm. If a central lattice point λ is labeled by an ordered K-tuple that has centroid $\tau \in \Lambda_{s/K}$, we say that λ is attracted by site τ. If two Voronoi cells $V_{s/K}(\tau_1)$ and $V_{s/K}(\tau_2)$ are spatially adjacent, we say that site τ_1 and site τ_2 are neighbors. In Fig. 8.8, site O and site T are neighbors, while site O and site M are not neighbors.

In Fig. 8.9, assume two neighboring sites a and b attract m and n central lattice points respectively. The m (n) central lattice points are labeled by m (n) nearest ordered

K-tuples centered at site a (b). For any point $x \in R^L$, let $J_{\overrightarrow{ab}}(x)$ be the projection value of x onto the axis $\overrightarrow{ab}$. Consider the set $S(a)$ of all the m points currently attracted by site a, and find

$$\lambda_{max} = \arg \max_{\lambda \in S(a)} J_{\overrightarrow{ab}}(\lambda). \tag{8.72}$$

Now, introduce an operator $\Gamma(a, b)$ that alters the label of λ_{max} to an ordered K-tuple of sublattice points centered at b. The effect of $\Gamma(a, b)$ is that sites a and b attract $m - 1$ and $n + 1$ central lattice points respectively, which are respectively labeled by $m - 1$ and $n + 1$ nearest ordered K-tuples centered at site a and site b.

From the definition of side distortion $d_s = \sum_{\lambda \in \Lambda} d(\lambda)P(\lambda)$ in (8.33), we have

$$d(\lambda) = \left(\frac{1}{K} \sum_{k=1}^{K} \|\alpha_k(\lambda) - \mu(\alpha(\lambda))\|^2 \right) + \left(\zeta \|\lambda - \mu(\alpha(\lambda))\|^2 \right). \tag{8.73}$$

Let us compute the change of $d(\lambda_{max})$ caused by the operation $\Gamma(a, b)$.

The change in the second term of $d(\lambda_{max})$ is

$$\zeta \left(\|\lambda_{max} - b\|^2 - \|\lambda_{max} - a\|^2 \right)$$
$$= \frac{\zeta}{L} \left(\left(J_{\overrightarrow{ab}}(\lambda_{max}) - L \|b - a\| \right)^2 - J_{\overrightarrow{ab}}(\lambda_{max})^2 \right). \tag{8.74}$$

Note the change of the second term is positive if $\lambda_{max} \in V_{s/K}(a)$.

The change in the first term is

$$f_b(n + 1) - f_a(m), \tag{8.75}$$

where $f_\tau(i)$ is the ith smallest value of $\frac{1}{K} \sum_{k=1}^{K} \|\lambda_k - \tau\|^2$ over all ordered K-tuples $(\lambda_1, \lambda_2, \cdots, \lambda_K) \in \Lambda_s^K$ such that $m(\lambda_1, \lambda_2, \cdots, \lambda_K) = \tau$.

The net change in $d(\lambda_{max})$ made by operation $\Gamma(a, b)$ is then

$$\Delta(a, b) = \zeta \left(\|\lambda_{max} - b\|^2 - \|\lambda_{max} - a\|^2 \right) + f_b(n + 1) - f_a(m). \tag{8.76}$$

If $\Delta(a, b) < 0$, then $\Gamma(a, b)$ improves index assignment.

The preceding discussions lead us to a simple local adjustment algorithm:

$(a^*, b^*) = \arg \min_a$ neighbors $_b \Delta(a, b)$;
While $\Delta(a^*, b^*) < 0$ do
    $\Gamma(a, b)$;
    $(a^*, b^*) = \arg \min_a$ neighbors $_b \Delta(a, b)$.

Note that it is only necessary to invoke the local adjustment $\Gamma(a, b)$ if the greedy algorithm does not simultaneously minimize the two terms of d_s.

Figure 8.10 shows the result of applying the local adjustment algorithm to the output of the greedy algorithm presented in Fig. 8.8. It is easy to prove that the local adjustment algorithm indeed finds the optimal index assignment for this case of three-description MDLVQ.

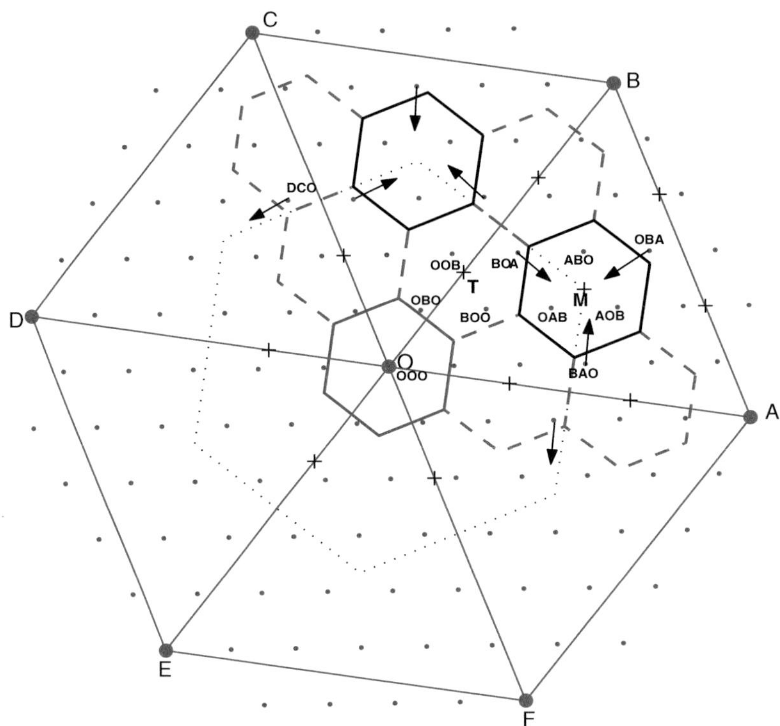

Figure 8.10 Optimal index assignments for the A_2 lattice, $N = 31, K = 3$. Points of Λ, Λ_s, and $\Lambda_{s/3}$ are marked by $\cdot$, $\bullet$, and $+$, respectively. © 2006 IEEE. Reprinted with permission from [132]

Finally, we conjecture that a combined use of the greedy algorithm and local adjustment $\Gamma^\rightarrow (a, b)$ solves the problem of optimal MDLVQ index assignment for any L-dimensional lattice and for all values of K and N.

8.6   PET-based MDC

PET-based MDC can be applied when we have a scalable source sequence. Assume that N is the number of desired descriptions and each description must contain L symbols (a symbol is a block of a fixed number of bits). In the PET framework, the source sequence is divided into L consecutive segments, and each of these segments is protected by RS code. Let m_i be the length (in symbols) of the i-th source segment, then the channel code assigned to protect the segment is the (N, m_i) RS code. The stream of these m_i source symbols followed by the $f_i = N - m_i$ redundancy symbols constitutes the i-th slice of the joint source-channel code. The effect of the (N, m_i) RS code associated with the i-th source segment is that, if at most f_i of N descriptions are lost, then all the m_i source symbols of the i-th slice can be correctly recovered. However, since the scalable source sequence is only sequentially refinable, decoding of the i-th source segment depends on

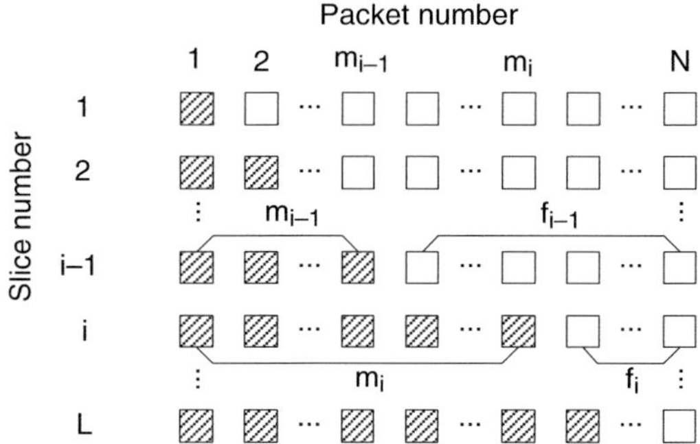

Figure 8.11 UEP packetization scheme. The dark rectangles correspond to source symbols and the white rectangles correspond to the redundancy symbols. © 2004 IEEE [76]

all the previous $i - 1$ segments, i.e., the complete prefix of the source sequence with respect to the current segment. Hence the number of redundancy symbols assigned to a slice must be monotonically non-increasing in the slice index:

$$f_1 \geq f_2 \geq \cdots \geq f_L. \tag{8.77}$$

The L-tuple $(f_1, f_2, \cdots, f_L)$ above is called L-slice redundancy assignment.

PET-based MDC is also referred to in the literature as uneven erasure protection (UEP) packetization (each description forms a packet to be transmitted over the lossy network) and we will use the latter terminology in the rest of this section.

Since no fractional protection symbols can be allocated in practice, we require that all f_i, $1 \leq i \leq L$ be integers between 0 and $N - 1$ and that the monotone relation (8.77) be enforced. Figure 8.11 illustrates the UEP packetization scheme. As we will see later, the constraint of decreasing redundancy level for subsequent segments of the source sequence can be lifted, if the rate-fidelity function of the scalable source sequence is convex. In this case, the solution of the unconstrained version of the optimization problem satisfies the constraint anyway. In practice, many scalable source codes are indeed constructed to shape an approximately convex operational rate-fidelity curve. A well-known example is the EBCOT technique used in JPEG 2000 [124].

Let $\phi(r)$ be the rate-fidelity function of the scalable source sequence, which is a monotonically non-decreasing function in rate $r \in [0, R_{max}]$, where R_{max} is the total number of source symbols. The efficiency of the redundancy assignment is measured by the expected fidelity of the reconstructed sequence at the decoder side. If exactly n packets are lost, such that $f_i \geq n > f_{i+1}$, for some i, then only the first i source segments can be completely recovered by the RS code. Since the source sequence is embedded, the receiver can decode it only up to the first lost symbol. Consequently, the receiver can decode only the first i source segments plus some few source symbols of the $(i + 1)$-th segment that are not lost. As in [89], [92], [89], [95], [38], we round off the effect of decoding these additional few symbols on fidelity. Hence the achieved fidelity is $\phi(r_i)$, where $r_i = \sum_{k=1}^{i} m_k = iN - \sum_{k=1}^{i} f_k$. The probability that the receiver achieves this fidelity is $\sum_{n=f_{i+1}+1}^{f_i} p_N(n)$, where $p_N(n)$ is the probability of losing n packets out of

N. Hence, the expected fidelity $\Phi(f_1, f_2, \cdots, f_L)$ of the reconstructed sequence at the decoder side can be expressed as

$$\Phi(f_1, f_2, \cdots, f_L) = \phi(0) \sum_{n=f_1+1}^{N} p_N(n) + \sum_{i=1}^{L-1} \phi(r_i) \sum_{n=f_{i+1}+1}^{f_i} p_N(n) + \phi(r_L) \sum_{n=0}^{f_L} p_N(n). \tag{8.78}$$

After straightforward algebraic manipulations we have:

$$\Phi(f_1, f_2, \cdots, f_L) = c_N(N)\phi(0) + \sum_{i=1}^{L} c_N(f_i)(\phi(r_i) - \phi(r_{i-1})), \tag{8.79}$$

where $c_N(k) = \sum_{n=0}^{k} p_N(n)$, $k = 0, 1, \cdots, N$, and $r_i = \sum_{k=1}^{i} m_k = iN - \sum_{k=1}^{i} f_k$, $1 \leq i \leq L, r_0 = 0$.

The objective of optimal UEP packetization is to find the redundancy assignment $(f_1, f_2, \cdots, f_L)$ that maximizes $\Phi(f_1, f_2, \cdots, f_L)$, for given N, L, $p_N(n)$, and $\phi(r)$.

Most of the existing algorithms proposed for this problem [89], [92], [89], [95] need the convexity of the rate-fidelity function to achieve optimality. Stockhammer and Buchner [38] present a dynamic programming algorithm of $O(N^2 L^2)$ time complexity that can obtain global optimality for convex rate-fidelity function. The algorithms of Puri and Ramchandran [92] and of Mohr, Riskin, and Ladner [89], [89] provide the globally optimal solution only if the rate-fidelity function is convex and fractional bit allocation is allowed. The algorithm of Stankovic, Hamzaoui, and Xiong [95] does not need the additional assumption of fractional bit allocation, but it can find only a local optimum. The only known globally optimal solution of UEP packetization problem in a general setting, with no assumptions on the rate-fidelity function or on the channel statistics, was given by Sachs, Anand, and Ramchandran [93]. It was a dynamic programming algorithm of $O(N^3 L^2)$ time complexity.

In [76], Dumitrescu, Wu, and Wang showed that the time complexity of the globally optimal UEP packetization can be reduced to $O(N^2 L^2)$ in the general case, and to $O(NL^2)$ if the rate-fidelity function of the scalable source is convex and if the probability $p_N(n)$ is monotonically non-increasing in n. Moreover, it was proved that the condition of monotone $p_N(n)$ can be removed for independent packet erasure channels with packet loss rate no larger than $\frac{N}{2(N+1)}$. In [79] further speed-up of the globally optimal solution is achieved in the convex case, under the same assumptions on the channel as stated above. The algorithm of [79] is based on a Lagrangian formulation of the problem. For each value of the Lagrangian multiplier λ, the algorithm requires $O(NL)$ time. The number of iterations needed to find the optimal λ, and hence to complete the algorithm, is much smaller than L, leading to great savings of computations from the $O(NL^2)$ time algorithm of [76].

The rest of this section presents the results of [76]. Precisely, the next subsection describes the $O(N^2 L^2)$ time globally optimal algorithm for the general case. We then proceed to the case of convex rate-fidelity function. First we show that the constraint of non-increasing redundancy assignment (8.77) can be lifted. Then we prove that the cost function underlying the optimal unconstrained UEP packetization problem has a strong monotone property called total monotonicity if the probability $p_N(n)$ is monotonically non-increasing in n. This property allows the use of a fast matrix search technique [71] to reduce the time complexity to $O(NL^2)$.

8.6.1 Exact solution for the general case

In this section we solve the optimal UEP packetization problem for general $\phi(r)$ and $p_N(n)$. Note that in (8.79) the term $c_N(N)\phi(0)$ is a constant and hence can be discarded in the optimal UEP design. In the summation that remains, the i-th term, $c_N(f_i)(\phi(r_i) - \phi(r_{i-1}))$, is the contribution of the i-th slice to the expected fidelity. The contribution of the first k slices to the expected fidelity, which depends only on the k-slice redundancy assignment $(f_1, f_2, \cdots, f_k)$, is denoted by $\Phi_k(f_1, f_2, \cdots, f_k)$:

$$\Phi_k(f_1, f_2, \cdots, f_k) = \sum_{i=1}^{k} c_N(f_i)(\phi(r_i) - \phi(r_{i-1})). \tag{8.80}$$

It is obvious that maximizing the expected fidelity $\Phi(f_1, f_2, \cdots, f_L)$ is equivalent to maximizing $\Phi_L(f_1, f_2, \cdots, f_L)$, which can be written as

$$\Phi_L(f_1, f_2, \cdots, f_L) = \Phi_k(f_1, f_2, \cdots, f_k) + C, \tag{8.81}$$

where $C = \sum_{i=k+1}^{L} c_N(f_i)(\phi(r_i) - \phi(r_{i-1}))$.

The algorithm development would be considerably simpler if the two terms $\Phi_k(f_1, f_2, \cdots, f_k)$ and C could be maximized separately. Unfortunately, the two expressions are not independent. C depends on the total number of protection symbols $f_1 + f_2 + \cdots + f_k$ for the first k slices (because $r_k = kN - (f_1 + f_2 + \cdots + f_k)$). Another dependency is imposed by the condition $f_k \geq f_{k+1}$ that has to be satisfied. We can break down these dependencies by fixing the total number of protection symbols t for the first k slices, $t = f_1 + f_2 + \cdots + f_k$, and by also fixing a value n satisfying $f_k \geq n \geq f_{k+1}$ (the value n has the significance of minimum redundancy required for the k-th slice and maximum redundancy admissible for the $(k+1)$-th slice). After introducing the new parameters n and t, consider the sub-problem of maximizing $\Phi_k(f_1, f_2, \cdots, f_k)$ over all k-slice redundancy assignments $(f_1, f_2, \cdots, f_k)$ for the first k slices, subject to $N - 1 \geq f_1 \geq \cdots \geq f_k \geq n$, $f_1 + f_2 + \cdots + f_k = t$. We call this sub-problem the sub-allocation (k, n, t). The sub-allocation (k, n, t) is defined for all triples (k, n, t) of integers such that $1 \leq k \leq L$, $0 \leq n \leq N - 1$ and

$$kn \leq t \leq k(N - 1). \tag{8.82}$$

For convenience of denotation, define

$$A(k, n, t) = \max \left\{ \Phi_k(f_1, f_2, \cdots, f_k) | N - 1 \geq f_1 \geq \cdots \geq f_k \geq n, \right.$$
$$\left. f_1 + f_2 + \cdots + f_k = t \right\}. \tag{8.83}$$

In order for such redundancy assignments to exist, the condition (8.82) has been imposed on the triple (k, n, t).

In the newly introduced notation the optimal UEP packetization problem can be restated as finding the L-slice redundancy assignment $(f_1, f_2, \cdots, f_L)$ such that

$$\Phi_L(f_1, f_2, \cdots, f_L) = \max_{0 \le t \le L(N-1)} A(L, 0, t). \tag{8.84}$$

In order to obtain the values $A(L, 0, t)$ for all possible t, we need to systematically evaluate $A(k, n, t)$ for all possible triples (k, n, t). The required computations are organized recursively as follows.

If the k-slice redundancy assignment $(f_1, f_2, \cdots, f_k)$ is optimal for the sub-allocation (k, n, t), then either $f_k = n$ or $f_k \ge n + 1$. If $f_k = n$ then the total number of protection symbols on the first $k - 1$ slices equals $t - n$ and the $(k - 1)$-slice redundancy assignment $(f_1, f_2, \cdots, f_{k-1})$ has to be optimal for the sub-allocation $(k-1, n, t-n)$, too. If $f_k \ge n+1$, then the assignment $(f_1, f_2, \cdots, f_k)$ has to be optimal for the sub-allocation $(k, n + 1, t)$ as well.

Note further that, for some triples (k, n, t), one of the two alternatives mentioned above ($f_k = n$ or $f_k \ge n + 1$) is impossible, hence only the other holds. This is the case when $kn \le t < k(n+1)$, which implies that f_k can not be larger than n; hence we obtain

$$A(k, n, t) = A(k - 1, n, t - n) + c_N(n)(\phi(kN - t) - \phi((k - 1)N - t + n)). \tag{8.85}$$

The other case happens when $n + (k - 1)(N - 1) < t \le k(N - 1)$, which implies that f_k can not be equal to n, and it follows that

$$A(k, n, t) = A(k, n + 1, t). \tag{8.86}$$

In all the other cases, i.e., for $k(n+1) \le t \le n+(k-1)(N-1)$, the following recursion holds:

$$\begin{aligned} A(k, n, t) &= \max\{A(k - 1, n, t - n) + c_N(n)(\phi(kN - t) \\ &\quad - \phi((k - 1)N - t + n)), A(k, n + 1, t)\}. \end{aligned} \tag{8.87}$$

The recursive formulae (8.85), (8.86), and (8.87) show that $A(k, n, t)$ can be computed in constant time provided that $A(k - 1, n, t - n)$ and $A(k, n + 1, t)$ are known. In order to take advantage of this result, we have to solve the sub-allocations (k, n, t) in such an order as to ensure that sub-allocation (k, n, t) is computed after sub-allocations $(k - 1, n, t - n)$ and $(k, n + 1, t)$. This can be done if k is enumerated in increasing order, but n in decreasing order, and for each given pair k, n, all possible t are considered before going to the next value of k or n.

After computing $\max_{0 \le t \le L(N-1)} A(L, 0, t)$, we need to restore the optimal L-slice redundancy assignment. Let $B(k, n, t)$ be the number of protection symbols for the k-th slice in the optimal k-slice redundancy assignment achieving $A(k, n, t)$ in (8.83), for each triple (k, n, t). We need to keep track of the values $B(k, n, t)$. To save space we do not store the actual values of $B(k, n, t)$ for each triple (k, n, t), but only a binary flag on which value is the maximum in the right-hand side of (8.87). Thus, we define $Z(k, n, t)$ to be 0 if (8.85) holds, and 1 otherwise, for all $1 \le k \le L$, $N - 1 \ge n \ge 0$, and $kn \le t \le k(N - 1)$. Upon computation of $A(k, n, t)$, the bit $Z(k, n, t)$ is also set and stored. The binary array $Z(\cdot, \cdot, \cdot)$ suffices to reconstruct any quantity $B(k, n, t)$. Namely,

$B(k, n, t) = n_1$, where n_1 is the smallest integer satisfying the conditions: $n \leq n_1 \leq t/k$, $Z(k, n_1, t) = 0$ and $Z(k, n', t) = 1$ for all $n', n \leq n' < n_1$. Such an integer n_1 always exists (indeed, for $n'' = \lfloor t/k \rfloor$ we have $Z(k, n'', t) = 0$). Hence, for a given triple (k, n, t), the reconstruction of $B(k, n, t)$ requires at most $O(N)$ time.

Summarizing the above, we present a UEP packetization algorithm to compute globally optimal L-slice redundancy assignment for arbitrary $p_N(n)$ and $\phi(r)$.

Algorithm A. Optimal UEP packetization for general case.

Step 1. For $k = 1, 2, \cdots, L$, $n = N - 1, N - 2, \cdots, 0$, and each t, $kn \leq t \leq k(N - 1)$, compute $A(k, n, t)$ and $Z(k, n, t)$ using recursion (8.85), (8.86), or (8.87), depending on t.

Step 2. Find $t_0 = \max_{0 \leq t \leq L(N-1)}^{-1} A(L, 0, t)$.

Step 3. Construct the optimal assignment $(f_1, f_2, \cdots, f_L)$ in the following way. Set $f_L = B(L, 0, t_0)$ and $t = t_0 - f_L$. For $k = L - 1, L - 2, \cdots, 1$, set $f_k = B(k, f_{k+1}, t)$ and then update t: $t = t - f_k$. The values $B(k, n, t)$ are determined as explained earlier.

A detailed pseudocode of the algorithm can be found in [76].

The time complexity of Step 1 is $O(N^2 L^2)$ because there are $O(N^2 L^2)$ entries of $A(k, n, t)$ and $Z(k, n, t)$, each of which is computed in $O(1)$ time. Step 2 clearly takes $O(NL)$ operations. Step 3 restores L quantities $B(k, n, t)$, each of which requires $O(N)$ time, taking $O(NL)$ time in all. Consequently, the overall time complexity of the algorithm is $O(N^2 L^2)$. The previously known algorithm for the same problem has a time complexity of $O(N^3 L^2)$ [93].

Next we analyze the space complexity. For each triple (k, n, t), $Z(k, n, t)$ has to be stored until the algorithm is completed. Since there are $O(N^2 L^2)$ such binary entries, $O(N^2 L^2)$ bits suffice to store $Z(\cdot, \cdot, \cdot)$. The values of $A(k, n, t)$ are not needed in the entire duration of Step 1, but only as long as they are used in (8.85), (8.86), or (8.87). Among the iterations on k, n, and t, k is the last one to vary. Therefore, after all quantities $A(k_0, n, t)$, for a fixed k_0, have been computed, the values $A(k, n, t)$ for $k < k_0$ and any n and t, are no longer needed and can be discarded. This means that only $O(N^2 L)$ entries of $A(k, n, t)$ need to be stored at any given time. Consequently, the space complexity of the algorithm is dominated by the memory requirement of $Z(\cdot, \cdot, \cdot)$, which is $O(N^2 L^2)$ in bits.

8.6.2 Fast matrix-search algorithm for convex case

The complexity of Algorithm A is still high. There are two ways to simplify Algorithm A and reduce its complexity. The first is to remove the constraint of non-increasing redundancy assignment to subsequent slices of the scalable source sequence in the optimization process, and hopefully without compromising optimality by doing so (this approach was taken by Stockhammer and Buchner [38], too). The second is to reduce the search space of the dynamic programming algorithm by the technique of fast matrix search [71]. This can be made possible if the underlying cost function satisfies a so-called total monotonicity. In this section we take these two steps to develop a more efficient algorithm for optimal UEP packetization.

The algorithm to be presented below finds the globally optimum of the UEP packetization problem in the case of convex rate-fidelity function, in a similar approach as the algorithm of [38] with $O(N^2 L^2)$ time complexity. But we go a step further and show that its time complexity can be reduced to $O(NL^2)$ under some mild assumptions about the channel statistics, namely if $p_N(n)$ is monotonically decreasing in n or if the channel is an independent erasure channel with packet loss rate ϵ no greater than $\frac{N}{2(N+1)}$.

The solution presented in the previous section for an optimal UEP packetization problem is found by recursively solving the sub-problems associated with parameters k, n, t. The parameter n was introduced to enforce the decreasing redundancy assignment, i.e., $f_1 \geq f_2 \geq \cdots \geq f_L$. If this condition is removed, then the parameter n can be dropped. In order to solve the problem of maximizing (8.79) without the constraint of (8.77), we recursively solve sub-problems associated with pairs of integers (k, t) with $1 \leq k \leq L, 0 \leq t \leq k(N-1)$. For each such pair (k, t) denote by $C(k, t)$ the analog of $A(k, 0, t)$ without the constraint (8.77). Namely,

$$C(k, t) = \max\{\Phi_k(f_1, \cdots, f_k) \mid 0 \leq f_1, \cdots, f_k \leq N - 1, f_1 + \cdots + f_k = t\}. \quad (8.88)$$

It is clear that $C(1, t) = \Phi_1(t)$ for all $t, 0 \leq t \leq N - 1$.

The computation of $C(k, t)$ can be done according to the recursive formula

$$C(k, t) = \max_{0 \leq t-j \leq N-1} (C(k - 1, j) + c_N(t - j)(\phi(kN - t) - \phi((k - 1)N - j))) \quad (8.89)$$

for all $2 \leq k \leq L, 0 \leq t \leq k(N-1)$.

Denote by $J(k, t)$ the largest value of j for which $C(k, t)$ is achieved in (8.89), for $2 \leq k \leq L, 0 \leq t \leq k(N-1)$, and set by convention $J(1, t) = 0$ for $0 \leq t \leq N - 1$. If there is more than one k-slice redundancy assignment $(f_1, \cdots, f_k)$ achieving $C(k, t)$ in (8.88), then we choose the one for which $f_1 + \cdots + f_{k-1} = J(k, t)$, hence for which the number of redundancy symbols on the k-th slice, f_k, is the smallest.

Consequently, the unconstrained version of the optimal UEP packetization problem can be solved by recursively computing the values $C(k, t)$ and $J(k, t)$ for all $1 \leq k \leq L$ and $0 \leq t \leq k(N-1)$. Then $t_0 = \max_{0 \leq t \leq L(N-1)}^{-1} C(L, t)$ is found and the optimal L-slice redundancy assignment for t_0 is reconstructed. We call this algorithm Algorithm B.

Next we state a proposition that governs the optimality of Algorithm B for the UEP packetization problem in the convex case. This result was proved in [76].

PROPOSITION 8.1   *If the rate-fidelity function $\phi(r)$ is convex, then the L-slice redundancy assignment computed by Algorithm B satisfies the constraint $f_1 \geq f_2 \geq \cdots \geq f_L$.*

Algorithm B computes $O(NL^2)$ instances of $C(k, t)$. If the computation of each value $C(k, t)$ is done by a full search, then it requires $O(N)$ time and the complexity of the algorithm becomes $O(N^2 L^2)$. However, one can do much better. This complexity can be reduced to $O(NL^2)$ by applying an elegant matrix-search technique introduced by Aggarwal *et al.* [71]. (A detailed description of this technique can be found in [73].) For each $k, 2 \leq k \leq L$, consider the upper triangular matrix G_k with the elements

$G_k(j,t), 0 \leq j \leq t \leq k(N-1)$, where $G_k(j,t)$ is defined by

$$G_k(j,t) = C(k-1,j) + c_N(t-j)(\phi(kN-t) - \phi((k-1)N-j)) \qquad (8.90)$$

for $t - N + 1 \leq j \leq t$, and $G_k(j,t) = -\infty$ for $0 \leq j < t - N + 1$. Then relation (8.89) becomes equivalent to

$$C(k,t) = \max_{0 \leq j \leq k(N-1)} G_k(j,t). \qquad (8.91)$$

In other words, for a given k, the search required by (8.89) corresponds to finding the maximum element in the column t of G_k. Matrix G_k is said to be totally monotone with respect to column maxima if for $j < j'$ and $t < t'$, the following implication holds:

$$G_k(j',t) \geq G_k(j,t) \Rightarrow G_k(j',t') \geq G_k(j,t'). \qquad (8.92)$$

As demonstrated by Aggarwal *et al.* in [70], all the m column maxima of an $m \times m$ matrix can be found in $O(m)$ time if the matrix is totally monotone. Very encouragingly, we can indeed show that the matrix G_k defined in (8.90) is totally monotone if $p_N(n)$ is monotonically non-increasing. This result is stated by the following proposition proved in [76].

PROPOSITION 8.2   *If the packet loss probability $p_N(n)$ is decreasing in n, then the upper triangular matrix G_k is totally monotone with respect to column maxima.*

The total monotonicity of G_k for non-increasing $p_N(n)$ enables us to apply the matrix-search technique of Aggarwal *et al.* [70] to compute all column maxima of G_k for given k, in $O(NL)$ time. Therefore, computing all values of $C(k,t)$ takes $O(NL^2)$ time. Consequently, the time complexity of optimal UEP packetization is reduced to $O(NL^2)$ when $\phi(r)$ is convex and $p_N(n)$ is non-increasing. In the case where the maximum value of column t is not unique, the matrix search algorithm chooses the one of the largest row index $j = J(k,t)$. This tie-break rule is the same as in Algorithm B, which is also a condition that we need in the proof of Proposition 8.1.

To assess the space complexity, we note that $O(NL)$ entries of $C(k,t)$ have to be stored at any given time for applying the recursion (8.89). Moreover, the algorithm needs to store $O(NL^2)$ entries of $J(k,t)$ in order to reconstruct the optimal redundancy assignment. The latter space requirement dominates, hence the overall space complexity is $O(NL^2)$.

The fast matrix search algorithm can also be applied to a wider class of erasure channels that do not even have non-increasing $p_N(n)$, as long as the rate-fidelity function is convex. Consider an independent erasure channel with the packet loss rate ϵ. The probability of losing n packets out of N is $p_N(n) = \binom{N}{n}\epsilon^n(1-\epsilon)^{N-n}$. The probability mass function $p_N(n)$ is not monotone, but it is a unimodal function with its peak at $n_0 = \lfloor \epsilon(N+1) \rfloor$. We have $p_N(n) \leq p_N(n+1)$ for $n < n_0$ and $p_N(n) \geq p_N(n+1)$ for $n \geq n_0$. Since $p_N(n)$ is decreasing for $n \geq n_0$, the fast matrix search technique can still be applied, if the algorithm restricts to redundancy assignments with at least n_0 protection symbols to each slice. In other words, we narrow the search range for j in (8.89) to

$$n_0 \leq t - j \leq N - 1. \tag{8.93}$$

According to Proposition 8.1, the assignment output by Algorithm B with this modification satisfies the constraint (8.77), too. Moreover, this assignment is still optimal for the UEP packetization problem, in spite of the restriction (8.93), if the packet loss rate ϵ is at most $\frac{N}{2(N+1)}$, as stated by the following proposition.

PROPOSITION 8.3   *If the rate-fidelity function $\phi(r)$ is convex and the channel is an independent erasure channel with packet loss rate $\epsilon \leq \frac{N}{2(N+1)}$, then there is an L-slice redundancy assignment maximizing (8.79) with at least n_0 protection symbols on each slice, where $n_0 = \lfloor \epsilon(N+1) \rfloor$.*

This result was proved in [76]. Note that the assumption $\epsilon \leq \frac{N}{2(N+1)}$ is very reasonable in practice since the threshold $\frac{N}{2(N+1)}$ is very close to the value 0.5.

The fast matrix search algorithm can also be used as an approximation algorithm when the rate-fidelity function $\phi(r)$ is not convex. We only need a slight modification to ensure that the output L-slice redundancy assignment satisfies the constraint (8.77). The modification is made in the matrix G_k for all k. The value $G_k(j, t)$ is set to -1 if $t - j > j - J(k - 1, j)$. In other words, the value j will not be a candidate at the selection of $J(k, t)$ if it introduces a reverse of the order in the final redundancy assignment. Note that this artificially imposed order may miss the globally optimal solution to the problem if $\phi(r)$ is not convex. But in practice, this seldom happens.

The algorithms proposed by Mohr, Riskin, and Ladner [89] and by Puri and Ramchandran [92] work on convex hull of operational rate-fidelity function $\phi(r)$ of the source sequence, and both can obtain optimal solution to the UEP packetization problem, if $\phi(r)$ is convex and if fractional redundancy bit allocation is allowed. Under the practical constraint of integer redundancy assignment, however, these two algorithms are still suboptimal even if $\phi(r)$ is strictly convex. The algorithm in [89] has the complexity $O(hN \log N)$, where h is the number of points on the convex hull of the rate-fidelity function. If $\phi(r)$ is convex, then its complexity becomes $O(N^2 L \log N)$, which is asymptotically higher than our $O(NL^2)$ algorithm, since typically $N \log N > L$ for reasonably long scalable sequences. The algorithm in [92] uses a Lagrangian multiplier λ. The algorithm is linear in the number of packets (N) for a given λ, and the value of λ to meet the target rate is found using a fast bisection search (it was usually found within 32 iterations according to [93]). The algorithm of Stankovic, Hamzaoui, and Xiong [95] is an $O(NL)$ time local search algorithm that starts from a solution that maximizes the expected number of source bits and iteratively improves the solution. It is the fastest among all UEP algorithms and offers very good approximation solution in practice.

8.7   General MDC based on progressive coding

The joint design framework in [125] involves diversity coding at the server node and routing and duplication at all other nodes. To achieve diversity coding at the servers, multiple-description coding (MDC) [47] is used for source coding. As advocated in [126], MDC is an efficient tool even in error-free networks. To simplify the joint design problem, the authors of [125] further propose the use of Balanced MDC (B-MDC). Unlike the general MDC where the quality of reconstruction depends on the particular subset of descriptions used in decoding, the reconstruction quality of B-MDC only depends on the total rate of the descriptions available at the decoder, and not on the actual subset of descriptions. This characteristic allows (as a suboptimal solution) breaking the problem in two parts: (a) optimal design of multicast trees used to transmit the descriptions such that the total rate of descriptions is received at all sinks to be maximized; (b) rate-distortion optimized B-MDC design given the multicast scheme found at step (a). For practicality, the B-MDC scheme based on Priority Encoding Transmission (PET) [72] is employed. The PET-based B-MDC framework [89] encodes consecutive layers of a scalable code stream using erasure protection codes of decreasing strength. The technique ensures that any subset of k descriptions recovers the first k data layers, leading thus to the same reconstruction quality irrespective of the particular subset of k descriptions received.

However, confining the MDC to be balanced may be too restrictive. The reason is that not all subsets of descriptions are necessarily received at network sink nodes. Specifically, if some subset S_1 of k descriptions is not received at any sink, then there is no need to ensure the same decoding quality for S_1 as for some other subset S_2 of k descriptions which is received at some sink. It is natural to think that by removing this constraint a better allocation of MDC rate between source and redundancy will be achieved, thus inherently improving the overall performance. Motivated by this observation, we extend the joint design framework to allow General MDC (G-MDC). Note that MDC in the general sense also includes successive refinement as a special case. Our G-MDC design approach is inspired by ideas used in linear network coding [48,127]. It uses linear coding to introduce redundancy, but achieves a more efficient use of the MDC budget by optimizing only subsets of descriptions which are received by the sinks.

The rest of the section is organized as follows. The target problem and the proposed G-MDC framework are described in 8.7.1. Then, 8.7.2 discusses the rate allocation in our new G-MDC scheme. The construction algorithm of the G-MDC is introduced in 8.7.3.

8.7.1   Framework description

Given a lossless directed acyclic network $G = \langle V, E \rangle$, we want to efficiently communicate a source content from the source node(s) to a set of sink nodes $T \in V$. The source is encoded using MDC and the descriptions are transmitted to the sinks using routing and duplication at intermediate nodes. The target problem is to jointly find the optimal MDC and routing scheme which maximize the overall reconstruction quality at all sinks.

Like in [125], we also solve the problem in two steps: (a) select the optimal multicast trees for all descriptions, (b) construct the general MDC such that the overall reconstruction fidelity is at all sinks to be maximized. In the first step, we use the same algorithm proposed in [125] to select the optimal multicast trees for the descriptions. This algorithm outputs a set of M multicast trees of equal rate, one corresponding to each description. Hence, all description packets will have the same size. For the second step, we propose an algorithm to construct the G-MDC that maximizes the overall reconstruction quality.

Not surprisingly, by removing the balanced constraint in the MDC design, we should find better, or at least equal, designs compared to B-MDC. However, finding optimal MDC is a prohibitive task. Therefore, we adopt a heuristic which is inspired by ideas developed in linear network coding. Like PET-based B-MDC, our G-MDC framework also assumes that the source is presented as a scalable code stream. In other words, any prefix of the code stream can be decoded ensuring some quality of the reconstruction, and, the longer the decoded prefix, the higher the reconstruction quality is. On the other hand, any symbol of the code stream can be decoded only if all previous symbols are also available. Also, like PET-based B-MDC, G-MDC divides the code stream in successive layers and ensures that any sink which receives k descriptions is able to recover the first k layers. However, as opposed to PET-based B-MDC, in G-MDC subsets of k descriptions, which are not received at any sink do not need to ensure decodability of the first k data layers. Therefore, G-MDC incorporates more flexibility in the redundancy assignment, thus leading to a more efficient allocation of the MDC rate between source and redundancy.

In our G-MDC framework, linear coding over a finite field F is used, where $|F| = 2^B$, $|F| > |T|$, T denoting the set of sinks. Each symbol is a sequence of B consecutive bits and is regarded as an element of this field. Each description symbol will be a linear combination of source symbols. Precisely, consider the partition of the sinks into subsets $T_1, \ldots, T_N$, such that all sinks in the same subset have the same flow value from incoming multicast trees; hence they receive the same number of descriptions. Moreover, the number of descriptions received at any sink in T_k is smaller than the number of descriptions arriving at any sink in T_{k+1}. Let $X_1, X_2, X_3, \cdots$ denote the successive source symbols. To build the G-MDC, the scalable source code stream is divided into N consecutive layers, the k-th layer containing $R_k - R_{k-1}$ source symbols, where $R_0 = 0$. Each description symbol s will carry a message $\mathbf{m}(s)$ which is a linear combination of data symbols, in other words

$$\mathbf{m}(s) = \sum_{j=1}^{R_N} c(s,j)X_j, \tag{8.94}$$

where $c(s,j) \in F$ for all s,j. Adopting the same terminology as in network coding, we call the R_N-dimensional vector $\mathbf{c}(s) = (c(s,j))_{1 \leq j \leq R_n}$ the encoding vector corresponding to symbol s. Moreover, the set of symbols forming each description is partitioned into N

levels. Let $L(s)$ denote the level assigned to symbol s. An essential condition in G-MDC is that any description symbol s has to carry a linear combination only of data symbols in the first $L(s)$ layers, in other words $c(s,j) = 0$ for all $L(s) + 1 \leq j \leq R_N$. To ensure that any sink $t \in T_k$ will be able to decode the first R_k data symbols, it is enough to guarantee that there is a set S_t of R_k description symbols received at sink t, which are linear combinations only of data symbols in the first k layers (hence S_t contains only symbols in the first k levels) and whose encoding vectors are linearly independent. Sink t uses only the messages carried by description symbols in S_t to recover the source, therefore these symbols will be called useful symbols for sink t. Precisely, if $s_1, \cdots, s_{R_k}$, denote the symbols in S_t, then the decoding at sink t is performed using the following linear operation:

$$[X_1, \cdots, X_{R_k}]^T = G_t^{-1}[\mathbf{m}(s_1), \cdots, \mathbf{m}(s_{R_k})]^T, \tag{8.95}$$

where G denotes the R_k-dimensional matrix whose i-th row consists of the first R_k components of $\mathbf{c}(s_i)$.

Note that in PET-based B-MDC, the symbols of all descriptions are also partitioned into levels. The k-th level is constructed by first dividing the k-th source layer into consecutive segments of k symbols and then applying to each segment a linear erasure code of codelength N, which corrects $N - k$ erasures. Therefore, we conclude that each description symbol in level k also carries a linear combination of source symbols, but only of source symbols in layer k. This constraint is relaxed in G-MDC where symbols in level k are allowed to carry linear combinations of the data symbols in the first k layers. Moreover, a B-MDC level is evenly divided between descriptions, while in G-MDC this condition is not imposed. Furthermore, in B-MDC it is required that every subset of k symbols from level k have independent encoding vectors, while in G-MDC the condition of encoding vectors independence is more relaxed. Based on the above observations it is clear that B-MDC has a higher flexibility in the redundancy assignment, and therefore will be able to achieve higher rate-distortion performance.

8.7.2   G-MDC rate optimization

The optimal G-MDC data rate allocation problem determines the number of symbols in each data layer, i.e., the values $R_1, \cdots R_N$, such that the average reconstruction quality at all sinks to be maximized. Also it assigns to each description symbol a level and determines the set S_t of useful symbols for each sink t. Note that while the level assigned to a symbol s is unique, the sinks for which s is useful can be more than one.

We assume that a fidelity function $\phi(R)$ is available, which represents the fidelity of the source reconstruction achieved after decoding a prefix of R symbols (e.g., negative distortion, or PSNR). Function $\phi(R)$ is increasing in R and upper convex (i.e., consecutive slopes are non-increasing: $\phi(R + 1) - \phi(R) \geq \phi(R + 2) - \phi(R + 1)$). The upper convexity assumption is a very good approximation in practice for multimedia source streams.

$$\max \qquad \sum_{k=1}^{N} \frac{|T_k|}{|T|} \phi(R_k) \tag{8.96a}$$

$$S.T.: \qquad R_1 \leq R_2 \leq \cdots \leq R_N \tag{8.96b}$$

$$R_j \geq \sum_{i \in \mathcal{D}_t} \sum_{l=1}^{j} x_i^{t,l}, \quad t \in T, 1 \leq j < L(t) \tag{8.96c}$$

$$R_k = \sum_{D_i \in \mathcal{D}_t} \sum_{l=1}^{k} x_i^{t,l}, \quad 1 \leq k \leq N, t \in T_k \tag{8.96d}$$

$$y_i^l = \max_{t \in T} \left\{ x_i^{t,l} \right\}, \quad 1 \leq i \leq M, 1 \leq l \leq N \tag{8.96e}$$

$$\sum_{l=1}^{N} y_i^l \leq C, \quad 1 \leq i \leq M \tag{8.96f}$$

$$x_i^{t,l} \geq 0 \quad 1 \leq i \leq M, 1 \leq l \leq N, t \in T \tag{8.96g}$$

$$x_i^{t,l} \quad are \quad integers \tag{8.96h}$$

Figure 8.12 Optimized redundancy allocation

Let C denote the size (i.e., number of symbols) for each description. For each sink t let $L(t)$ denote the unique value k such that $t \in T_k$. Also denote the descriptions by $D_1, \cdots, D_M$, and for each sink t let $\mathcal{D}_t$ be the set of descriptions received at t.

The problem of optimal rate allocation can be cast as an integer programming problem with upper convex objective function and linear constraints. For this we introduce the following variables. For each sink t, level l, $1 \leq l \leq L(t)$, and description D_i received at sink t, let $x_i^{t,l}$ denote the number of symbols in level l of description D_i which are useful for sink t. Moreover, let y_i^l denote the total number of symbols in level l of description i. Then the optimization problem is formulated as in Fig. 8.12. The objective function represents the average fidelity of the reconstruction at all sinks. Condition (8.96d) follows from the fact that the size of S_t equals $R_{L(t)}$ and all symbols of S_t must be included in the first $L(t)$ levels. Condition (8.96c) ensures that the number of symbols in S_t which are linear combinations only of data symbols in the first j source layers, is not larger than the size of the first j source layers.

Note that, according to the discussion in the previous section, the PET-based B-MDC scheme also satisfies the constraints in Fig. 8.12, and additionally some more. Therefore, we conclude that our G-MDC scheme guarantees at least the same performance as PET-based B-MDC. It is true that the proposed G-MDC design needs more information about the network compared to B-MDC. However, this is not a disadvantage of G-MDC, at least in our scenario, since this additional information, precisely, the knowledge of the multicast trees, is already available after the step (a) of the network transmission problem was completed.

Since the cost function in Fig. 8.12 is upper convex, the optimization problem can be further simplified by linearizing the objective function. For this we use the standard method for linearization when the cost is an upper convex piecewise linear function. Consider integers $0 = n_0 < n_1 < n_2 < \cdots < n_q = R_{max}$ such that $\phi(\cdot)$ is piecewise linear on each set $\{n_j + 1, \cdots, n_{j+1}\}$. In other words, there are real values $c_1, \cdots, c_q$

such that, for any j and R with $n_j < R \leq n_{j+1}$, we have

$$\phi(R) = \phi(0) + \sum_{i=1}^{j} c_i(n_i - n_{i-1}) + c_{j+1}(R - n_j). \tag{8.97}$$

Upper convexity of $\phi(\cdot)$ ensures that $c_1 > c_2 > \cdots > c_q$. Note that the values n_i, c_i that satisfy the above conditions are guaranteed to exist. In the worst case, i.e., with the largest number q of piecewise linear portions, we have $n_i = i$ and $c_i = \phi(i) - \phi(i-1)$.

Further, consider the additional integer variables $r_{k,j}, 1 \leq j \leq q, 1 \leq k \leq N$ satisfying the constraints

$$0 \leq r_{k,j} \leq n_j - n_{j-1}, \; 1 \leq j \leq q, 1 \leq k \leq N \tag{8.98}$$

$$\sum_{j=1}^{q} r_{k,j} = R_k \; 1 \leq k \leq N \tag{8.99}$$

and the cost function

$$\sum_{k=1}^{N} \frac{|T_k|}{|T|} \sum_{j=1}^{q} c_j r_{k,j}. \tag{8.100}$$

Clearly, if $n_{j_0} < R_k \leq n_{j_0+1}$, then (8.100) is maximized under the constraints of (8.98) and (8.99), when $r_{k,j} = n_j - n_{j-1}$ for $j < j_0$ and $r_{k,j} = 0$ for $j > j_0$. In this case the value of (8.100) equals the value of the cost function in (8.96a). Therefore, the optimization problem in Fig. 8.12 can be transformed to a linear integer programming problem by replacing the cost function with (8.100) and adding the constraints of (8.98) and (8.99). Further, the new integer linear programming problem can be solved by using approximation techniques based on linear programming relaxation. Moreover, to reduce the computational complexity, the value of q can be kept small by using an approximation of the fidelity function with a smaller number of piecewise linear portions.

8.7.3 G-MDC description construction

After solving the above problem, the descriptions are divided into levels based on the values y_i^l. Then the useful symbols for each sink t are determined as follows. For each i and l the first $x_i^{t,l}$ symbols in level l of descriptions D_i are marked as useful symbols for each sink t satisfying $1 \leq l \leq L(t)$ and $D_i \in \mathcal{D}_t$.

To complete the description construction, i.e., to assign the message $\mathbf{m}(s)$ to each description symbol s, we need to determine the encoding vector for each symbol s which is useful for some sink. This algorithm is described in Fig. 8.13.

The algorithm initially assigns to each description symbol s several temporary encoding vectors, one such vector, denoted $\mathbf{c}(s,t)$, corresponding to each sink $t \in \mathcal{T}_s$, where $\mathcal{T}_s$ denotes the set of sinks for which s is a useful symbol. Let $W_t = \{\mathbf{c}(s,t)|s \in S_t\}$. In the second round, the algorithm assigns to each symbol s the final encoding vector $\mathbf{c}(s)$, and sets $\mathbf{c}(s,t) = \mathbf{c}(s)$ for all $t \in \mathcal{T}_s$. The algorithm maintains the invariant that for each sink t, the set W_t contains $R(L(t))$ linearly independent vectors, and each vector in W_t corresponding to some symbol s has the last $R_N - R_{L(s)}$ components equal to 0.

The correctness of Algorithm 1 follows from the following lemmas.

> **for** *each $t \in T$* **do**
> > sort S_t in ascending order of symbols level
> > **for** *each $s \in S_t$* **do**
> > > $\mathbf{c}(s, t) = [0^{j-1}, 1, 0^{R_N - j}]$, where j is the position of s in the sorted list S_t
> >
> > **end**
> >
> **end**
> **for** *each description D_i* **do**
> > **for** *each symbol s of D_i* **do**
> > > Choose $\mathbf{v} \in F^{R_N}$ such that $\mathbf{v}$ is a linear combination of $\{\mathbf{c}(s, t') | t' \in \mathcal{T}_s\}$ and $\forall t \in \mathcal{T}_s$,
> > > $\mathbf{v}$ is linearly independent of $W_t \setminus \{\mathbf{c}(s, t)\}$;   (*)
> > > $\mathbf{c}(s) = \mathbf{v}$;
> > > **for** *each $t \in \mathcal{T}_s$* **do**
> > > > $\mathbf{c}(s, t) = \mathbf{c}(s)$
> > >
> > > **end**
> > >
> > **end**
> >
> **end**

Algorithm 1: Construction of G-MDC

Figure 8.13 Construction of G-MDC

LEMMA 8.5   *After assigning the temporary encoding vectors in the first round, the invariant holds for all $t \in T$.*

Proof   Let us fix arbitrarily some sink t. Condition (8.96d) ensures that S_t contains exactly $R_{L(t)}$ symbols, hence W_t contains $R_{L(t)}$ vectors. It is obvious that these vectors are linearly independent.

Moreover, because of the constraint (8.96c), the total number of symbols in S_t situated in levels 1 through i is at most R_i, for any $1 \leq i \leq L(t)$. Therefore, if j is the position of some symbol s in the sorted list S_t, then clearly we have $j \leq R_{L(s)}$, which implies that the last $R_N - R_{L(s)}$ components of vector $\mathbf{c}(s, t)$ are equal to 0.   □

LEMMA 8.6   *The encoding vector $\mathbf{v}$ in step $(*)$ can be found, when $|F| > |T|$.*

Proof   This proof closely follows the proof of Lemma 4 in [48]. The R_N-dimensional vector $\mathbf{v}$ must satisfy

$$\mathbf{v} = \sum_{t' \in \mathcal{T}_s} u_{t'}(s)\mathbf{c}(s, t'). \tag{8.101}$$

for $u_{t'}(s) \in F$. Let us call the vector $[u_{t'}(s) : t' \in \mathcal{T}_s]$ local encoding vector. Thus, we have to show that there exists a local encoding vector such that $\mathbf{v}$ is linearly independent of $W_t \setminus \{\mathbf{c}(s, t)\}$ for any $t \in \mathcal{T}_s$. Since all $\mathbf{c}(s, t)$ have the last $R_N - R_{L(s)}$ components equal to 0, (8.101) will further ensure that $\mathbf{v}$ has the same property.

Let us fix some $t \in \mathcal{T}_s$ and for each $t' \in \mathcal{T}_s$ denote by $\mathbf{c}'(s, t')$ the $R_{L(t)}$-dimensional vector obtained from $\mathbf{c}(s, t')$ after removing the last $R_N - R_{L(t)}$ components. Since $L(s) \leq L(t)$, it follows that all removed components are 0. Define $\mathbf{v}'$ in the same manner. Then relation (8.101) implies

$$\mathbf{v}' = \sum_{t' \in \mathcal{T}_s} u_{t'}(s)\mathbf{c}'(s, t'). \tag{8.102}$$

Now for each $s' \in S_t$ denote by $\mathbf{c}'(s', t)$ the $R_{L(t)}$-dimensional vector obtained from $\mathbf{c}(s', t)$ after removing the last $R_N - R_{L(t)}$ components. All the removed components are 0, since $L(s') \leq L(t)$. Due to the invariant, the set of vectors $\{\mathbf{c}'(s', t) : s' \in S_t\}$ forms a basis of $F^{R_{L(t)}}$. Therefore, $\mathbf{v}'$ can be written as a linear combination of the vectors in this basis

$$\mathbf{v}' = \sum_{s' \in S_t, s' \neq s} \beta(s')\mathbf{c}'(s', t) + (\beta + u_t(s))\mathbf{c}'(s, t) \tag{8.103}$$

where β is uniquely determined and it does not depend on $u_t(s)$. Clearly, $\mathbf{v}$ is linearly dependent on $W_t \setminus \{\mathbf{c}(s, t)\}$ if and only if $u_t(s) = \beta$. It follows that for any choice of the set $\{u_{t'}(s) | t' \in \mathcal{T}_s, t' \neq t\}$, only one choice of $u_t(s)$ makes $\mathbf{v}$ linearly dependent on $W_t \setminus \{\mathbf{c}(s, t)\}$. We conclude that there are $|F|^{|\mathcal{T}_s|-1}$ invalid local encoding vectors for a receiver $t \in \mathcal{T}_s$, and the total number of invalid local encoding vectors is at most $|T| \cdot |F|^{|\mathcal{T}_s|-1} < |F|^{|\mathcal{T}_s|}$. Therefore, there must exist at least one valid local coding vector. $\square$

LEMMA 8.7 *Any sink $t \in T$, is guaranteed to decode the first $R_{L(t)}$ source symbols.*

Proof By Lemmas 8.1 and 8.2 the invariant holds at the end of the algorithm. Hence t receives $R_{L(t)}$ messages carried by the useful symbols, which are linear combinations of data symbols $X_1, \cdots, X_{R_{L(t)}}$. By the invariant the encoding vectors corresponding to symbols in S_t are linearly independent, a fact which ensures that all $X_1, \cdots, X_{R_{L(t)}}$ can be decoded. $\square$

Constructing S_t and W_t for all $t \in T$ takes $O(|T|R_N)$ time. Note that the sets S_t can be constructed such that they are already sorted in ascending order of symbols level. The encoding vector $\mathbf{v}$ at step (*) can be found in $O(|T|^2 R_N)$ time similarly to the deterministic implementation in [48]. Therefore, finding the encoding vectors for all useful symbols requires $O(CM|T|^2 R_N)$ time. We conclude that the total running time of Algorithm 1 is $O(CM|T|^2 R_N)$.

9 Using progressive codes for lossy source communication

In the previous chapters, we studied the merits of using balanced multiple-description codes, along with carefully optimized routing strategies for lossy source communication in heterogeneous networks. As discussed in the first part of this book, network coding is able to improve the network communication throughput, when compared to routing only. Starting from this chapter, we consider the merits of using network coding for lossy source communication. The problem becomes very complex in its most general form. Some of the complexities are apparent from the discussion in the following section. A practical subclass of the problem uses progressive source codes along with carefully optimized network coding strategies for efficient multicast of compressible sources, and is the subject of this chapter. But first, some general notes on using network coding for lossy communications.

9.1 Lossy source communication with network coding: an introduction

The lossy source communication strategies studied so far have been based on routing of multiple-description codes. The descriptions have been routed and duplicated in the network, without network coding. In this chapter, we start introducing lossy source communication methods that may use network coding ideas for network delivery. We start by introducing a generalization of the network coding problem, called Rainbow Network Coding (RNC). Unlike the current formulation of network coding, RNC recognizes the fact that the information communicated to different members of a multicast group can be different in general.

Take the scenario depicted in Fig. 9.1 as an example. Here, two information bits are to be communicated from node 1 to the sink nodes. The capacity of all links is one.

First, let's assume that nodes 4,5 are sinks. By Theorem 2.1, two bits a, b can be communicated to nodes 5,6 simultaneously, as indicated in Fig. 9.1(a). Here, transcoding at relay nodes plays a crucial role. In particular, node 7 combines the two bits it receives to produce $a \oplus b$ (where $\oplus$ indicates XOR operation) which enables nodes 4,5 to recover both bits. It is easy to verify that at most 3 bits can be communicated to the nodes 4,5 using routing only (see discussion in Section 2.2). Therefore, the multicast throughput with routing only is at most 1.5 bits per second per node, when the set of sink nodes is $T = \{4, 5\}$ (see Fig. 9.1(b)).

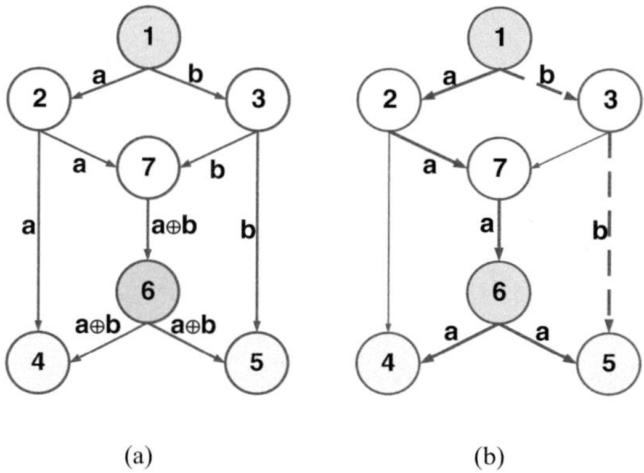

Figure 9.1 Multicasting common information: (a) With network coding (the XOR operation at node 7), two bits of information can be sent to nodes 4,5. This is not possible without network coding. Note that node 6 does not receive either of the two bits. (b) With routing only, however, two bits can be sent to node 5, one bit to each of the two nodes 4,6

Now, what if the set of sink nodes is $T = \{4, 5, 6\}$? Since the min-cut (and hence max-flow) into node 6 is only one, the multicast capacity of the network with sinks $\{4, 5, 6\}$ is only one. Therefore, only one bit of *common information* per second can be communicated to each sink. Both examples in Fig. 9.1, however, are able to communicate a total of 4 information bits to nodes 4,5,6. In Fig. 9.1(a), nodes 4,5 each receive 2 bits, while node 6 does not receive any (a total of 4 bits). In contrast, in Fig. 9.1(b), nodes 4,6 each receive one bit, while node 5 receives two bits (again, a total of 4 bits). Note that in this case, a different set of bits is available to each of the three members of the multicast group.

If each atomic data entity for transmission is assigned a unique "color," the above problem is about delivering to each node with as large a "spectrum" of colors as possible, without requiring all sink nodes to have exactly the same set of colors. Due to this analogy, we call this form of network information flow problem Rainbow Network Coding (RNC), a generalization of Rainbow Network Flow (RNF) to the case where network coding is allowed.

Both examples in Fig. 9.1(a) and Fig. 9.1(b) are valid rainbow network codes. In Fig. 9.1(a), the RNC involves transcoding of information at the relay node 7 (the XOR operation). The example in Fig. 9.1(b), however, uses routing only. This subclass of RNC (called RNF) was our main communication strategy in Chapter 5.

In this chapter, we introduce the idea of layered multicast (LM). In LM, progressive or layered source codes are used for source compression, while network coding is used to multicast these coding layers. The receivers that receive more layers are able to reconstruct the source at lower distortions. The multicast strategies, however, should respect the layered structure of the codes, i.e., a layer should be delivered to a receiver, only if the receiver has received all the previous layers. We start by formulating the problem of LM.

9.2   Formulation

The problem of layered multicast (LM) is defined on a directed acyclic multi-graph $\mathcal{G}\langle V, E \rangle$, a single-server node $s \in V$, and a set of clients $T \subset V$. Each network link e has capacity $C(e)$ that indicates the average number of bits that can be communicated over e without error, per network use. For any integer n, the multi-graph $G\langle V, E \rangle$ is created by replacing each link of $\mathcal{G}\langle V, E \rangle$ with $\lfloor n \times C(e) \rfloor$ links, each with capacity 1. A total of K information messages called *layers* are generated at s of lengths $h_1 = n \cdot r_1, h_2 = n \cdot r_2, \ldots, h_K = n \cdot r_K$ for some $r_k \in \mathbb{R}^+, k = 1, 2, \ldots, K$. We will assume a fixed large integer n is given such that h_k's can be approximated by integers, and we deal with $G_n\langle V, E \rangle$ only. The dependence of the parameters on the block size n is understood throughout. The use of integer link capacity is a well-known technique to simplify graph theoretical arguments and does not restrict the generality of the derivations.

DEFINITION 9.1 (multicast session and multicast strategy)   A multicast session is identified by a tuple $\Gamma_k(h_k) = (G_k, h_k)$, where G_k is a subnetwork (not necessarily a sub-tree) of G. A multicast strategy Γ is an ordered collection of K multicast sessions, $\Gamma = (\Gamma_1(h_1), \Gamma_2(h_2), \ldots, \Gamma_K(h_K))$, such that each edge e belongs to at most one multicast session.

Multicast session k is used to multicast layer k of length h_k bits. Define $V_k \subset V$ as the set of all nodes in G_i that are able to recover all the h_k bits in layer k without error.

DEFINITION 9.2 (subscription)   A client node t *may subscribe* to a multicast session k if $t \in V_k$. Define $T_k \subset V_k \cap T$ as the set of all clients that subscribe to layer k. A subscription strategy Γ is said to be layered if and only if for all $T_k, k = 2, 3, \ldots, K$ one has $T_k \subset T_{k-1}$.

DEFINITION 9.3 (layered flow vector)   The total number of bits of layered messages received by node t is defined as $q_t = \sum_{k:t \in T_k} h_k$. The vector $(q_t; t \in T)$ is called the *layered flow vector*.

DEFINITION 9.4 (achievable layered flow vector)   Define $Q \in \mathbb{R}^{+|T|}$ as the union of all layered flow vectors $(q_t; t \in T)$ for all layered multicast strategies Γ.

Let $(f_t; t \in T)$ be the value of max-flow to the sink nodes in T. The maximum number of information bits that can flow from s to t is at most f_t bits per network use [8].

DEFINITION 9.5 (inter-layer vs. intra-layer network coding)   If network coding is applied to the communication of each session separately, the layered coding strategy is called "intra-layer" coding. On the other hand, if the layers in different sessions are encoded together, the layered coding strategy is called "inter-layer" coding.

Note that the set of achievable flow vectors using intra-layer network coding is a subset of that achievable with inter-layer network coding.

DEFINITION 9.6 (absolute optimality)   Since f_t is the max-flow into t, we must have $q_t \leq f_t$ for all $t \in T$. If $(f_t; t \in T) \in Q$, i.e., if there exists a layered multicast strategy that can deliver the maximum possible flow to each client, we say that layered multicast is absolutely optimal and the corresponding layered multicast strategy is called the absolutely optimal layered multicast strategy.

Remark 9.1   If $f_{min} = \min_{t \in T} f_t$ then $(f_{min}; t \in T) \in Q$, i.e., a flow vector where all components are equal to f_{min} is always achievable. In particular, an absolutely optimal layered multicast strategy with one multicast session always exists if all the sink nodes have the same max-flow.

9.3   Layered multicast with intra-layer network coding

In multi-rate multicast problems where clients have different max-flows, absolutely optimal multicast strategies, as defined in the previous section, may not exist. We start by conditions under which an absolutely optimal, intra-layer multicast strategy exists for a given network and how that strategy may be efficiently constructed. We then move to optimization strategies for intra-layer coding in general networks. In the next section, we will investigate optimized intra-layer coding strategies.

9.3.1   Conditions for absolute optimality of LM with intra-layer network coding

The following theorems show the special cases where LM with intra-layer network coding is absolutely optimal.

THEOREM 9.1   *An absolutely optimal LM strategy with at most two layers always exists if the number of sink nodes is at most two. Counter-examples with three clients can be found for which no absolutely optimal LM strategy exists.*

Proof   If the sink nodes have the same max-flow (which includes the case with only one sink as a trivial special case), the absolutely optimal LM strategy has a single multicast session which consists of the set of maximum edge disjoint paths to the receivers. We only need to prove the theorem for the case of two sink nodes with max-flows $f_1 < f_2$ without loss of generality. We know that there are a maximum of f_i edge disjoint paths to the sink nodes i, for $i = 1, 2$. Let $A = \{a_1, a_2, \ldots, a_{f_1}\}$ and $B = \{b_1, b_2, \ldots, b_{f_2}\}$ be two sets of maximum edge disjoint paths into the sink nodes 1 and 2 respectively. Define $B_\cap = \{b_1, b_2, \ldots, b_{f_1}\}$ and $B_* = \{b_{f_1+1}, b_{f_1+2}, \ldots, b_{f_2}\}$.

Suppose for a moment that the f_1 paths in A are edge-disjoint from the paths in B_*. In this case, we can construct the multicast strategy with two multicast sessions as follows: the first layer consists of the union of paths in $B_\cap$ and A, i.e., $\Gamma_1 = B_\cap \bigcup A$ of size f_1. The second layer is $\Gamma_2 = B_*$ and has rate $f_2 - f_1$. Node 2 subscribes to both layers and receives data of size $f_1 + (f_2 - f_1) = f_2$, while the first node subscribes only to the first layer and receives data of size only f_1. The strategy is therefore absolutely optimal.

There remains to show that for any $f_1 < f_2$, there exists a set of edge-disjoint paths $A = \{a_1, a_2, \ldots, a_{f_1}\}$ to node 1 and $B = \{b_1, b_2, \ldots, b_{f_1}, b_{f_1+1}, \ldots, b_{f_2}\}$ to node 2, such that $\{b_{f_1+1}, b_{f_1+2}, \ldots, b_{f_2}\} \cap A = \phi$. Our proof repeats part of the argument for a network decomposition suggested in [4] which in turn is inspired by the ideas in [67].

Let B, A be sets of maximum edge-disjoint paths from s to t_1 and t_2 respectively. Thus, we have $f_1 = |B| \geq f_2 = |A|$. Color each edge in B and A either Red, Green or (Red,Green). If an edge belongs only to B, it is colored pure Green. If it belongs only to A, it is colored pure Red, and if it belongs to both A and B it is colored (Red,Green). If there are at least $f_2 - f_1$ purely Green paths in B, then we are done. If not, we can re-color the paths as follows:

Take a path P in B that starts from s and ends in t_1, and find e, the first edge whose color is (Red,Green). If this edge is the first edge in P, then stop. A (Red,Green) edge means that there exists exactly one path P' in A that contains edge e. Now color all Green edges in P from s to e (including e itself) (Red,Green). For all edges in path P', from s to e (but not including e itself), remove the color Red from the edge. Repeat this procedure until either there is no edge with color (Red,Green) in P or only the first edge in P is colored (Red,Green).

Repeat the above procedure for all the paths P in B. Once the process is done, we can claim that there is at least $|B| - |A| = f_2 - f_1$ purely Green paths from s to t_2. To see this, first note that when the algorithm stops, if an edge e belonging to a path P in B is Green, then all the edges in P after e must be purely Green. Now one can show that there are at least $f_2 = f_1$ purely Green edges that go out of s. This is true because whenever the algorithm removes color Red from an outgoing edge of s, it will place Red on another outgoing edge. Thus, the number of outgoing edges of s that have color Red remains constant and is equal to f_1. This means that at least $f_2 - f_1$ outgoing edges of s must be purely Green, which means that there are at least $f_2 - f_1$ purely Green paths from s to t_2. The multicast strategy will route the first multicast session of rate f_1 over the rest of the edges to both t_1 and t_2. Sink t_2 further subscribes to the second multicast session consisting of the $f_2 - f_1$ purely Green paths to t_2 and each sink node receives information at a rate equal to its max-flow. $\qquad\square$

Remark 9.2 This result should be compared with those in [62] where the layered multicast constraint does not exist and sink nodes are allowed to receive any subset of the information messages, in which case, authors in [62] show that information at a rate equal to max-flow can be delivered to any three clients in the network.

The next result is concerned with the existence of absolutely optimal LM strategies for more general problems. An acyclic network induces a (not necessarily unique) topological order on the nodes. For two nodes $v, v' \in V$, we say $v \prec v'$ if a directed path from v to v' exists. An ordering of $G\langle V, E \rangle$ is an ordered sequence $(v_1, v_2, \ldots, v_{|V|})$ of all nodes in V, such that $v_i \prec v_j$ for all $1 \leq i < j \leq |V|$. An ordering of V, therefore, specifies a hierarchical topological structure on G.

An ordering of V is said to be flow-consistent with respect to the set of client T, if for all $t_i, t_j \in T$ such that $t_i \prec t_j$ we have $f_{t_i} \geq f_{t_j}$ (remember that f_u is the max-flow to the node u). In other words, *sink nodes* higher up in the hierarchy should have larger max-flows. One expects this property to hold in distribution networks with a top-down structure. It turns out that an absolutely optimal layered multicast strategy always exists for such networks:

THEOREM 9.2 *Layered multicast is absolutely optimal if $G\langle V,E\rangle$ has a max-flow consistent topological ordering with respect to T.*

Before moving to the proof, we need to establish the following lemma.

LEMMA 9.1 *Suppose a common message of length h is available at nodes $S = \{v_1, v_2, \ldots, v_M\} \subset V$. Take N other nodes $U = \{u_1, u_2, \ldots, u_N\} \subset V$. For each $u_i, i = 1, 2, \ldots, N$, assume that there exist h edge disjoint paths, each of capacity 1, that start from one of the M nodes in S and end at u_i. Then, the message h can be multicast to all nodes in U from the nodes in S, without error, using network coding.*

Proof This is a generalization of the network coding in multicast setting with one source node, i.e., when $S = \{s\}$. To prove this, let's add a dummy node v to the network, which we call the super source, and assume that the h bits message is also available at v. Next, add h virtual links, each of capacity 1, between v and each of the nodes in S. Clearly, there are at least h edge-disjoint paths from v to each node in U. Thus, by results in [8], the h bits of message can be multicast from v to all N nodes in U using network coding. Take one such network coding strategy and consider the information bits communicated from v over each of the $h \times M$ virtual links. Each of these information bits must be a deterministic function of h information bits available to v. But all these bits are also available to every node in S. Thus, a node in S can calculate all the h bits that it receives from v, on its own. This means that one could remove the super source v (and all the virtual links), and still be able to multicast the h-bit information message to the nodes in U. $\qquad\square$

Proof Define P_t as the set of all edge-disjoint paths from s to t. Take one of these paths p, that starts from s and ends at $t \in T$, but passes through client nodes $t_0 = s, t_1, t_2, \ldots, t_m$ along the way (we have defined $t_0 = s$ for consistency). Define $\mathcal{P}(p)$ as the portion of p that starts from the last client node in this chain before t, i.e., t_m, and ends at t. Also, define $\mathcal{A}(p)$ as the portion of p from $\mathcal{P}(p)$ to t.

Let $f_1 < f_2 < \ldots < f_K = f_{max}$ be the sorted set of max-flow values of all clients. We choose as $T_k \subset T$ the set of all clients with max-flow f_k and multicast layers to their intended clients, starting by the uppermost hierarchy (T_K). There are f_{max} edge disjoint paths into each client $t \in T_{max}$. By definition of max-flow consistency, none of the paths in $P_t, t \in T_K$ passes through a sink node. By using network coding, we can communicate all the K layers to T_K, with total rate of $h_1 + h_2 + \ldots h_K = f_1 + (f_2 - f_1) + \ldots + (f_K - f_{K-1}) = f_K = f_{max}$. Thus, all nodes in T_K are fully satisfied.

We continue our proof by induction. Assume that all the sink nodes in T_{K-i} for $i = 0, 1, \ldots, n$ are fully satisfied, that is, they receive information of size f_{K-i}. We now prove that the nodes in T_{K-n-1} can also receive information of length f_{K-n-1} and are hence fully satisfied. With our induction assumption, all nodes in S_t have at least the

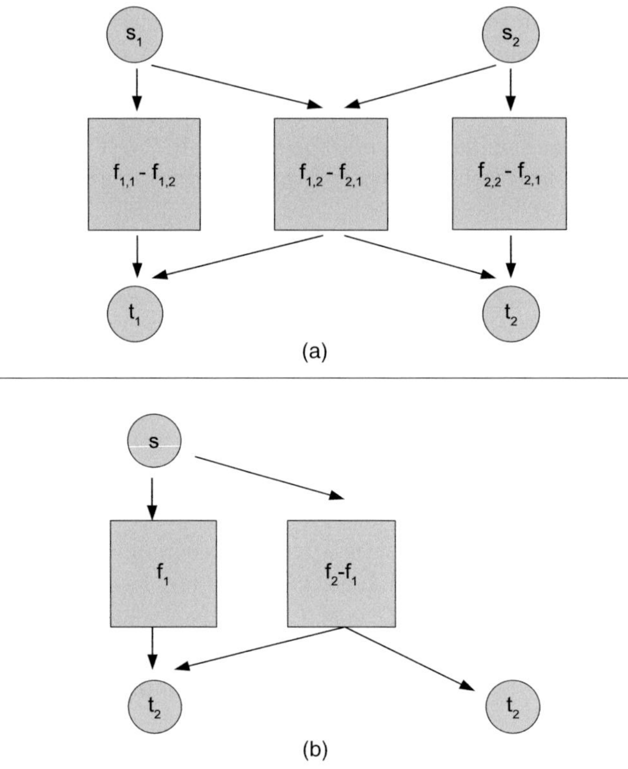

Figure 9.2 (a) The separation idea of Ramamorthy *et al.* [4] and (b) its specialization to the proof of our theorem

first n layers of information. We will multicast the first $n-1$ layers to nodes in T_{K-n-1}. Take a $t \in T_{K-n-1}$. For all $p \in P_t$, the paths $\mathcal{A}(p)$ are clearly disjoint. Since there are f_{K-n-1} edge-disjoint paths to $t \in T_{K-1}$, we must have $|\cup_{p\in P_t} \mathcal{A}(p)| = f_{K-1}$.

Define the set $S_t = \cup_{p\in P_t}\mathcal{P}(p)$. Since the network topology is max-flow consistent, we must have $S_t \subset T_K \cup T_{K-1} \cup \ldots \cup T_{K-n} + s$. Therefore, all nodes in S_t have access to the first n layers of information. Now, note that for each $p \in P_t$, the path $\mathcal{A}(p)$ has a root in one of the nodes in S_t. Therefore, for each $t \in T_{K-n-1}$, there exists f_{K-n-1} edge-disjoint paths, each rooted in one of the nodes in S_t. Therefore, the first $n-1$ layers receive information of total size f_{K-n-1}. Thus, we have proved that all nodes in T_{K-n-1} are also fully satisfied. Which completes the proof. $\qquad\square$

9.3.2   Optimization of layered multicast strategy

Results in the previous subsection showed that for some special networks, intra-layer LM with network coding is absolutely optimal. The work of [49] analyzes the problem of designing efficient layered multicast strategies for general directed networks, when only intra-layer network coding is allowed. The authors tackle a practical optimization problem in this framework.

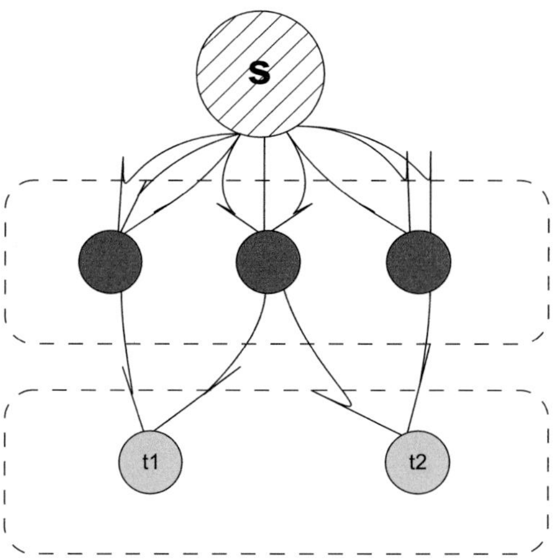

Figure 9.3 The schematics of the proof of Theorem 9.2: Each node in an upper hierarchy can act as a source for a subset of the layers for nodes in the lower hierarchies

Problem: Find a subscription strategy with K sessions (where K is also a variable) and find a feasible flow vector $[q_t; t \in T]$ such that the total rate of layered multicast, i.e., $R = \sum_{t \in T} q_t$.

Note that there are two questions that have to be answered: first, the partitioning of the receivers into K partitions, and then finding the best feasible flow vector. The two problems have to be optimized jointly, in general. However, the joint optimization problem appears to be hard. On the other hand, if the partitioning of the receivers is fixed, the optimization of the flow rates can be performed in polynomial time, as shown in [49].

To partition the receivers, the authors propose the following: partition the receivers in T into K groups such that all the receivers in a given group have the same max-flow. Formally, let $\hat{f}_1, \hat{f}_2, \ldots, \hat{f}_K$, for $K \leq |T|$, be the sorted sequence of the max-flows of receivers in T. Then, the partitions are defined as $T_k = \{t \in T : f_t = \hat{f}_k\}, k = 1, 2, \ldots, K$. While this partitioning is not necessarily optimal, it ensures a form of weak fairness, i.e., it ensures that the total flow to a node with a larger max-flow is never less than that of a node with a lower max-flow. Experiments on random networks also confirm the effectiveness of this choice.

The proposed solution in [49] is based on the following six steps:

- Partition the set of receivers T into subsets $T_1, T_2, \ldots, T_K$ such that: (i) the receivers in each subset T_k have identical max-flows, and (ii) the max-flow of receivers in T_k is greater than the max-flow of those in T_{k-1}, for $k > 1$.
- Construct K graphs, $G_1, G_2, \ldots, G_K$, such that $G_j = \langle V_j, E_j \rangle$ and $V_j = V - \cup_{i=1:j-1} T_i$ and $E_j = \{(u, v) \in E : u, v \in V_j\}$.

$$\max \quad \sum_{i=1}^{K} q_i \tag{9.1a}$$

$$\text{subject to} \quad \sum_{(s,v)\in E_j} f_{i(s,v)j} = q_j - q_j - 1 \quad \forall i \in (T \cap V_j), \forall 1 \leq j \leq K \tag{9.1b}$$

$$\sum_{(u,i)\in E_j} f_{i(u,i)j} = q_j - q_{j-1} \quad \forall i \in (T \cap V_j), \forall 1 \leq j \leq K \tag{9.1c}$$

$$\sum_{u\in In_j(v)} f_{i(u,v)j} = \sum_{u\in Out_j(v)} f_{i(u,v)j} \quad \forall v \in V_j, 1 \leq j \leq K \tag{9.1d}$$

$$Y_{(u,v)j} \geq f_{i(u,v)} \quad \forall i \in T \cap V_j, 1 \leq j \leq K \tag{9.1e}$$

$$\sum_{j=1}^{x(u,v)} Y_{(u,v)j} \leq C_{(u,v)}, \quad x(u,v) = \max l : (u,v) \in E_l \tag{9.1f}$$

Figure 9.4 Intra-layer flow optimization

- Construct and solve the linear programming problem in Fig. 9.4 to determine $q_t k$, for $k = 1, 2, \ldots, K$, which are the rates delivered to receivers in T_k, and parameters $Y_{(u,v)j}$ where $(u, v) \in E_j$ for $1 \leq j \leq K$. In this formulation, $f_{i(u,v)j}$ is the rate allocated to receiver i in subgraph G_j on edge (u, v), and $Y_{(u,v)j}$ is the total capacity allocated to all receivers in subgraph G_j on edge (u, v).

  The constraints in Fig. 9.4 can be explained as follows. Constraint (9.1b) is the flow-balance constraint at the source. Constraint (9.1c) corresponds to flow-balance constraints at the receivers. Similarly, constraint (9.1d) gives the flow-balance constraints for all intermediate nodes in the graph. For each node, the total outgoing flow rate is equal to the total incoming flow rate on each path to a receiver. Constraint (9.1e) is the network coding constraint and (9.1f) is the edge capacity constraint.
- Augment the graphs $G_1, G_2, \ldots, G_K$ with edge weights as determined from the linear programming problem. In particular, edge (u, v) in graph G_j is assigned a weight $Y_{(u,v)j}$.
- Using the Ford-Fulkerson max-flow algorithm, determine paths to receivers in G_j, for all $j = 1, 2, \ldots, K$, such that each receiver in T_k receives a flow of $q_k - q_{k-1}$.
- Use the code construction algorithm in [11] to deliver a net flow of q_k to all receivers in T_k.

The methods above allowed only for intra-layer network coding, i.e., network coding was allowed only within a given layer, and the data of the two layers are not mixed. Better results are possible if one allows for inter-layer network coding, as discussed in the next section.

9.4   Layered multicast with inter-layer network coding

Consider source segments (or messages) $x_1, \ldots, x_M$, of equal size, and a directed acyclic network $G = (V, E)$ (the edge capacities in the given network are normalized by the size of a source segment), a set of source nodes S, and a set of sink nodes T. All data messages $x_1, \ldots, x_M$ are available at all source nodes. A transmission scheme using network coding allows each message sent along an edge to be a function of the messages received at the node where the edge originates. We define a layered multicast code achieving rate $R(t)$ at each sink t, as a transmission scheme using network coding which ensures that each sink t can recover the first $R(t)$ source segments after decoding the received messages. We assume a non-decreasing fidelity function $\phi(R)$ is given, representing the fidelity of the reconstruction after decoding the first R messages. The simplest example is $\phi(R) = R$. Other examples of fidelity functions, meaningful in multimedia applications, are PSNR, SNR, or the negative distortion. The problem we address in this work is designing a layered multicast code such that $\sum_{t \in T} \phi(R(t))$ is maximized.

To make clearer the "layered" characteristic, we give an equivalent formulation to this problem. We partition the set of data sequences into several layers, and each sink subscribes a certain number of layers. Note that, due to the property of scalable source coding, only the base layer and the following consecutive layers can contribute to the reconstruction fidelity. By grouping together the sink nodes which receive the same data flow layers, we get a partition of sink nodes $T_1, \ldots, T_N$, such that $R(t) = R(t')$ for any t, t' in the same subset T_k, and $R(t) < R(t')$ for any $t \in T_k$, $t' \in T_{k+1}$ and any k. Let R_k denote the common value of $R(t)$ for the sinks $t \in T_k$. Define the k-th data layer as the set of source messages $\{x_{R_k+1}, \ldots, x_{R_{k+1}}\}$. Clearly, the layered multicast code guarantees that any sink in T_k receives the first k data layers. Then the above problem can be reformulated as follows.

PROBLEM 9.1   Find the partition $\mathcal{T} = \{T_1, \ldots, T_N\}$ (where N is also a variable), the values $0 < R_1 < R_2 < \cdots < R_N$, and a layered multicast code achieving rate R_k at each sink $t \in T_k$, for each k, such that $\sum_{k=1}^{N} \sum_{t \in T_k} \phi(R_k)$ is maximized.

Without loss of generality, the following development is confined to the case of a single source node. The case of multiple sources can be converted into one with a single source by adding a super source node and connecting the super source node to all source nodes by edges of infinite capacity. This conversion is depicted in Fig. 9.5.

9.4.1   Flow optimization for inter-layer network coding

9.4.1.1   Preliminaries

Solving Problem 9.1 over all possible sink partitions is a difficult task. A simplification of the problem is to impose a fixed partition $\mathcal{T}$ of the sinks and find the optimal rate allocation corresponding to $\mathcal{T}$. Intuitively, the number of layers received by each sink should be proportional to the sink's max-flow value. This intuition motivates choosing for $\mathcal{T}$ the partition induced by the max-flow values, i.e., where all sinks in the same

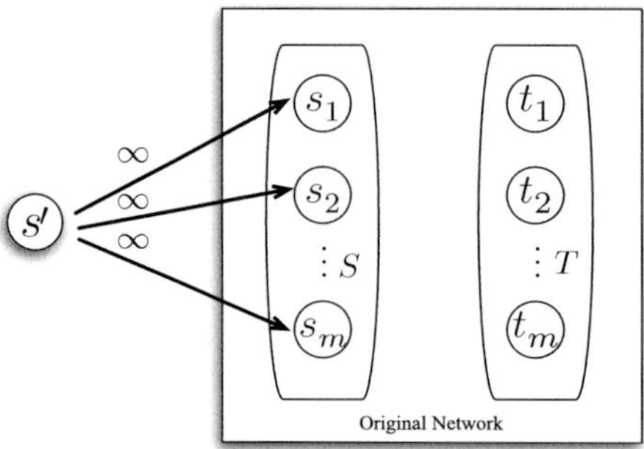

Figure 9.5 Converting a problem with multiple sources to one with a single source. © 2009 IEEE. Reprinted with permission from [143]

subset have the same max-flow value and moreover, the max-flow value of any sink in T_k is smaller than the max-flow value of any sink in T_{k+1}. We will denote this partition by $\mathcal{T}_{\text{max-flow}}$. The partition $\mathcal{T}_{\text{max-flow}}$ was used in prior works [49,50]. Therefore, imposing this constraint will not lead to loss in performance versus the techniques in [49, 50].

Having the partition of sinks specified, our target problem can be formulated as follows.

PROBLEM 9.2   Given a partition $\mathcal{T} = \{T_1, T_2, \ldots, T_N\}$ of the set of sinks, find the values $0 \leq R_1 \leq R_2 \leq \cdots \leq R_N$ and a layered multicast code achieving rate R_k at each sink $t \in T_k$, such that $\sum_{k=1}^{N} \sum_{t \in T_k} \phi(R_k)$ is maximized.

Remark 9.3   By allowing equality between consecutive rates R_k and R_{k+1} in the formulation of Problem 9.2, we actually perform a search over all sink partitions obtained from $\mathcal{T}$ by cumulating consecutive subsets. (The equality $R_k = R_{k+1}$ means that T_k and T_{k+1} are merged into a single subset.)

As discussed in the previous sections, some layered multicast formulations using network coding [49–51, 138] perform intra-layer network coding only. Specifically, each layer of data is transmitted to the sinks in a multicast layer. Network coding is applied only inside each layer, not across layers. Data flow in different multicast layers cannot be encoded together, i.e., the messages transmitted in multicast layer k can only be the linear combination of source segments in data layer k.

In order to understand the drawback of the intra-layer technique and to provide some insights into the merits of the inter-layer scheme, we analyze the example shown in Fig. 9.6. The network illustrated in the figure is a unit capacity network with 4 sinks t_1, t_2, t_3, t_4. The max-flows of the sinks are 2, 2, 2, 1, respectively. In order to achieve the max-flow rates at all sinks, the source data have to be divided into two layers x_1 and x_2. x_1 has to be delivered to all sinks in the first multicast layer, and x_2 must be delivered to sinks t_1, t_2, t_3 in the second multicast layer. Since cooperation across layers is not allowed, no edge can transmit the combination of x_1 and x_2. Thus, the optimal layered multicast solution shown in Fig. 9.6 (which in this case is the optimal multi-rate multicast solution as well) cannot be achieved.

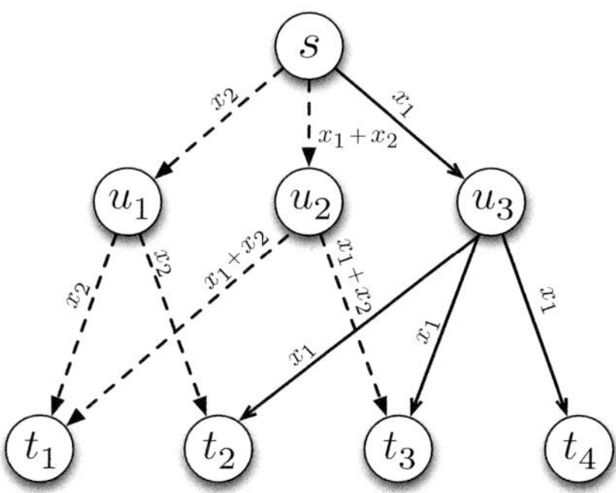

Figure 9.6 Optimal layered multicast solution for the given network. © 2009 IEEE. Reprinted with permission from [143]

The above example was first used in [52] when discussing the multilevel diversity coding problem. The author proved that the optimal rate cannot be achieved if network coding across layers was not allowed.

By carefully examining the optimal solution we observe that the flow can indeed be divided into two multicast layers, but using different criteria than in the intra-layer formulations. Consider the first multicast layer to consist of edges depicted in Fig. 9.6 with solid lines and the second multicast layer to consist of dotted line edges. It can be seen that the second multicast layer carries data in both data layers, not just in data layer 2. Precisely, edges (s, u_2), (u_2, t_1), (u_2, t_3) transmit a combination of x_1 and x_2. Moreover, note that sink t_1 does not receive any unit of flow in the first multicast layer, while in the intra-layer schemes it would receive one unit of flow. But then in order to compensate, t_1 receives two units of flow in the second multicast layer, while under the intra-layer framework it would receive only one. The above observations could be interpreted in the following way: for some sinks, part of the data in the first layer is "delayed" and transmitted together with the data in the second layer.

This thought leads to a novel layered multicast technique based on inter-layer network coding. As usual, the transmission is divided into multicast layers, but the concept of a multicast layer in our framework is significantly different from the inter-layer network coding formulations. Precisely, the flow in the k-th multicast layer is not confined to carry only combinations of data in data layer k. Instead, it is allowed to transport combinations of all data in the first k data layers. This is how network coding across data layers is performed. Another notable difference is that the amount of flow delivered to different sinks during a single multicast layer is not necessarily the same. In other words, each multicast layer is not unirate. The number of data layers received by each sink is decided by the sink partition. Sinks in subset T_k will receive k layers of data. Moreover, we allow flow to transfer from one layer to a higher layer. The transfer of flow at the source node could be explained in the following way. In the first multicast layer the source node s has available R_1 units of flow (i.e., data layer 1) for transmission to any sink. If only $R_1 - 1$ units of flow are transmitted to some sink $t \in T_k$, then the remaining unit of flow is available for transmission to sink t in the second multicast layer, together with the new $R_2 - R_1$ flow units corresponding to data layer 2. This unit of flow could be used in multicast layer 2 (i.e., transmitted to t), in which case we say that it was transferred to layer 2, or it can be "delayed" and transmitted to t in a higher layer. However, by the "end" of multicast layer k all delayed flow must reach sink t in order to ensure that t receives all data in the first k data layers. Flow transfer from a lower multicast layer to a higher layer is admitted at intermediate nodes as well.

Now the proposed solution to Problem 9.2 consists of two major parts. First a flow optimization problem is solved. Next a single network code is constructed to achieve the optimal rates for all sinks.

9.4.1.2   Inter-layer flow optimization

For every node $v \in V$, let $In(v)$ denote the set of incoming links to v and let $Out(v)$ denote the set of outgoing links from v. $C_{i,j}$ is the capacity of edge (i,j), and s is the source node.

We divide the flow into N layers. Any sink $t \in T_k$, for some k, may receive flow in the first k layers with the requirement that the total amount of flow received over the first k layers is to equal R_k. Let $x_{i,j}^{t,l}$ be the flow on edge (i,j) for sink t in layer l. Define $b_j^{t,l}$ to be the potential of node j for sink t in layer l. Negative node potential indicates a supplying node, while positive node potential indicates a demanding node. Let $y_{i,j}^k$ be the actual flow on edge (i,j) in layer k (over all sinks). For each sink node t, let $L(t)$ denote the number of data layers that sink t will receive (i.e., $L(t) = k$ if and only if $t \in T_k$). Then, the flow optimization problem can be formulated as shown in Fig. 9.7.

Constraint (9.2c) follows from the definition of node potential. (9.2d), (9.2e) are the potential constraints at the source node. The total flow sent out from the source s to sink t over the first $L(t)$ layers should equal $R_{L(t)}$. But the flow sent to t over the first j layers, $j < L(t)$, can be less than R_j because some part of the flow in lower layers can be "delayed" and transmitted in higher layers.

Constraints (9.2f), (9.2g) concern the potential of sink nodes. Since flow in lower layers can transfer to higher layers, the total flow received by sink t over the first j layers, $j < L(t)$, can be less than the total flow sent out to t by the source s over those layers. But the total flow received over the first $L(t)$ layers should equal the total flow sent by the source over those layers.

(9.2h), (9.2i) are the constraints for the potentials at the intermediate nodes. Note that the potential of some intermediate node l can be negative for some layers. For example, negative potential in layer j means that some flow in layer $1, \ldots, j-1$, transfers to layer j at the current node. Since the transfer of flow is allowed only from a lower to a higher layer, the total incoming flow (designated to sink t) at intermediate node l, over the first j layers, cannot be smaller than the total outgoing flow. Moreover, if $j = L(t)$, the two quantities have to be equal.

Since the amount of flow which transfers from layer $1, \ldots, j-1$ to layer j should be less than or equal to the sum of flow in layer $1, \ldots, j-1$, the sum of potentials in the first j layers must be non-negative.

Constraint (9.2j) is the network coding constraint, meaning that the flow for different sinks in the same layer can be combined together. Constraint (9.2k) confines that the actual flow in each edge cannot exceed the edge capacity.

$$\max \quad \sum_{t \in T} \phi(R_{L(t)}) \tag{9.2a}$$

$$\text{subject to} \quad R_1 \leq R_2 \leq \cdots \leq R_N \tag{9.2b}$$

$$b_j^{t,l} = \sum_{(i,j) \in In(j)} x_{i,j}^{t,l} - \sum_{(j,h) \in Out(j)} x_{j,h}^{t,l} \tag{9.2c}$$

$$R_j \geq -\sum_{i=1}^{j} b_s^{t,i}, \quad \forall t \in T, \quad 1 \leq j < L(t) \tag{9.2d}$$

$$R_j = -\sum_{i=1}^{j} b_s^{t,i}, \quad \forall t \in T, \quad j = L(t) \tag{9.2e}$$

$$\sum_{i=1}^{j} b_t^{t,i} \leq -\sum_{i=1}^{j} b_s^{t,i}, \quad \forall t \in T, \quad 1 \leq j < L(t) \tag{9.2f}$$

$$\sum_{i=1}^{j} b_t^{t,i} = -\sum_{i=1}^{j} b_s^{t,i}, \quad \forall t \in T, \quad j = L(t) \tag{9.2g}$$

$$\sum_{i=1}^{j} b_n^{t,i} \geq 0, \quad \forall t \in T, \quad 1 \leq j < L(t), \quad n \notin \{s,t\} \tag{9.2h}$$

$$\sum_{i=1}^{j} b_n^{t,i} = 0, \quad \forall t \in T, \quad j = L(t), \quad n \notin \{s,t\} \tag{9.2i}$$

$$y_{i,j}^{l} = \max_{t \in T} \left\{ x_{i,j}^{t,l} \right\}, \quad \forall l \quad 1 \leq l \leq N \tag{9.2j}$$

$$\sum_{l=1}^{N} y_{i,j}^{l} \leq C_{i,j}, \quad \forall (i,j) \in E \tag{9.2k}$$

$$x_{i,j}^{t,l} \text{ is non-negative integer}, \quad \forall t \in T, \quad \forall (i,j) \in E, \quad \forall l \quad 1 \leq l \leq N \tag{9.2l}$$

Figure 9.7 Inter-layer flow optimization

Finally, notice that inequality (9.2b) follows from (9.2e) since the sinks potentials are non-negative, thus (9.2b) can be removed.

9.4.1.3 Example

The following example illustrates a solution to the above flow optimization problem and the proposed intra-layer network code. To better illustrate the idea, we use a linear cost function $\phi(R) = R$. Consider the unit capacity network shown in Fig. 9.8. s is the source node and t_1, t_2, t_3 are the sinks. The max-flow to t_1, t_2, t_3 are 2, 2, 1 respectively, and therefore, sinks are divided into two subsets: $T_1 = \{t_3\}$, $T_2 = \{t_1, t_2\}$. The flow paths found by the flow optimization algorithm (indicated by the values $y_{i,j}^{k}$) are shown in Fig. 9.9. Edges in the first layer are shown as solid lines, while edges in the second layer are shown as dashed lines.

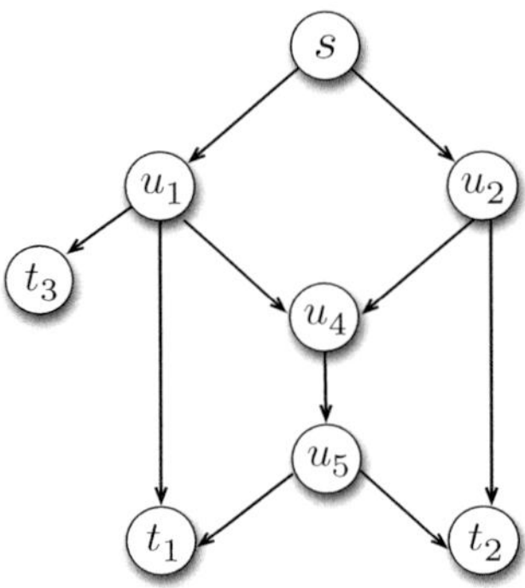

Figure 9.8 Example network with 3 sinks. © 2009 IEEE. Reprinted with permission from [143]

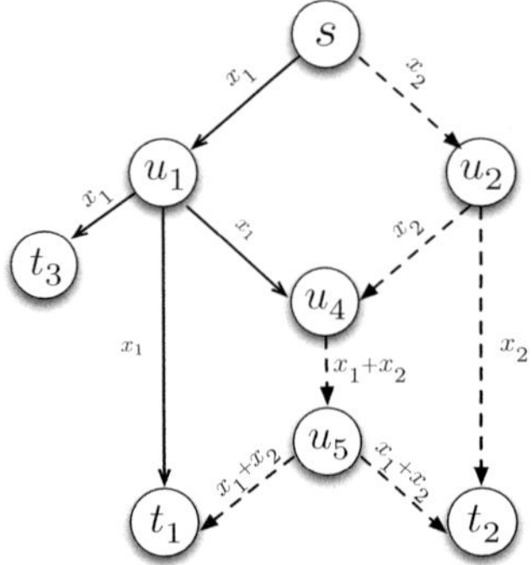

Figure 9.9 Solution produced by the intra-layer network coding scheme. This is the optimal layered multicast solution for the given network. © 2009 IEEE. Reprinted with permission from [143]

Note that under the proposed inter-layer network coding framework, in order to achieve the optimal solution, we would need two multicast layers, as shown in Fig. 9.10. Edge (u_4, u_5) should be included in both multicast layers in order to carry x_1 to t_2 and x_2 to t_1. This is not possible since the edge is a unit capacity edge. On the other side, in the inter-layer network coding framework, edge (u_4, u_5) is included in the second multicast layer and carries $x_1 + x_2$. Then the information about x_2 reaches sink t_1 via the path $s - u_2 - u_4 - u_5 - t_1$, which is completely included in layer 2. The information about data x_1 reaches sink t_2 via the path $s - u_1 - u_4 - u_5 - t_2$, which has the first two edges in the first layer and the next two edges in the second layer. This path illustrates the concept of transfer of flow between layers. At node u_4, one unit of flow directed to sink t_2 transfers from layer 1 to layer 2.

9.4.1.4 Observations

An important observation is that the solution of the intra-layer network coding flow optimization problem, when the sinks partition is $\mathcal{T}_{\text{max-flow}}$, is guaranteed to be at least as good as that of intra-layer network coding flow optimization schemes [49,50]. This is due to the fact that the intra-layered optimization formulation is included in the proposed

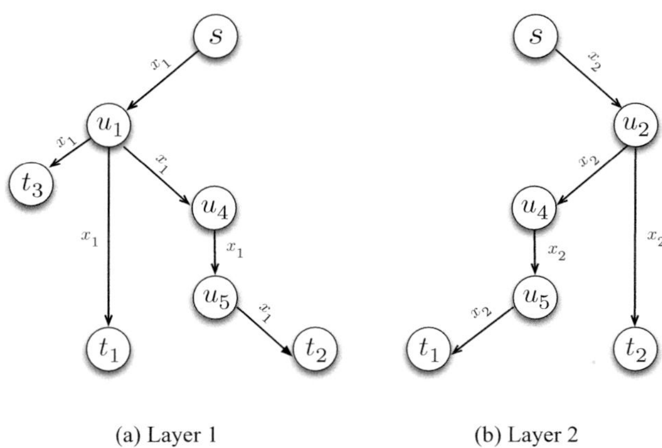

Figure 9.10 Optimal layered multicast solution for network in Fig. 9.8 cannot be achieved with the intra-layer network coding technique unless edge (u_4, u_5) has capacity 2. © 2009 IEEE. Reprinted with permission from [143]

inter-layered formulation as a special case. Precisely, if we change the inequalities in constraints (9.2d), (9.2f), (9.2h) into equalities, we will obtain the exact formulation of layered multicast in [50].

Another notable observation is that the solution of the problem in Fig. 9.7 improves as the partition $\mathcal{T}$ becomes finer. In order to justify this claim, consider two partitions $\mathcal{T}_1 = \{T_1, \ldots, T_N\}$ and $\mathcal{T}_1' = \{T_1', \ldots, T_{N'}'\}$, such that $\mathcal{T}_1$ is finer than $\mathcal{T}_1'$. Then there are integers $1 = m_1 < m_2 < \cdots < m_{N'} < m_{N'+1} = N + 1$ such that $T_k' = \cup_{j=m_k}^{m_{k+1}-1} T_j$, $1 \le k \le N'$. Then any feasible solution of flow optimization problem corresponding to $\mathcal{T}_1'$ can be converted to a feasible solution corresponding to $\mathcal{T}_1$, by letting the flow in multicast layer m_k (on each edge and for each sink) for the latter case be equal with the flow in multicast layer k for the former case, $1 \le k \le N'$, and by assigning zero flow in any other multicast layer for $\mathcal{T}_1$. Therefore, in order to improve the performance of the proposed layered multicast scheme, we will consider a $|T|$-size sinks partition in Problem 9.2, in other words, a partition where each subset consists of only one sink.

9.4.1.5   Linearization of flow optimization problem

A simple choice for the fidelity function is $\phi(R) = R$. Then the problem of Fig. 9.7 is a linear integer program. However, there are other more meaningful fidelity measures for multimedia applications (e.g., PSNR, SNR, negative mean squared error), and the solution which maximizes the overall received flow is not necessarily the solution with the highest overall reconstruction fidelity. Thus, in order to improve the performance of the layered multicast scheme, the real fidelity function (which is not linear) is more suitable. To handle such a case, we will convert the flow optimization problem into a linear integer program.

Let the fidelity function be a non-decreasing function $\phi(R)$, defined for any integer $R, 1 \le R \le M$. Recall that M is the number of source segments available at the source node s, for transmission during a time unit. Because function $\phi(\cdot)$ is non-decreasing, it

$$\max \quad \sum_{k=1}^{N}\sum_{j=1}^{M} c_j r_{k,j} \tag{9.5a}$$

$$\text{such that} \quad \sum_{j=1}^{M} r_{k,j} = R_k, \quad 1 \le k \le N \tag{9.5b}$$

$$r_{k,1} \ge r_{k,2} \ge \cdots \ge r_{k,M}, \quad 1 \le k \le N \tag{9.5c}$$

$$r_{k,j} \in \{0,1\}, \quad 1 \le k \le N, \quad 1 \le j \le M \tag{9.5d}$$

$$\text{conditions (9.2b)–(9.2l) hold} \tag{9.5e}$$

Figure 9.11 Linearization of the flow optimization problem

follows that there are non-negative real numbers c_j, $1 \le j \le M$, such that

$$\phi(R) = \phi(0) + \sum_{j=1}^{R} c_j \times 1 + \sum_{j=R+1}^{M} c_j \times 0, \tag{9.3}$$

for any integer R with $1 \le R \le M$. To linearize the flow optimization problem, we introduce the additional binary variables $r_{k,j} \in \{0,1\}$, $1 \le k \le N$, $1 \le j \le M$. The value of $r_{k,j}$ indicates whether or not the data segment x_j is included in the first k data layers. Precisely, $r_{k,j} = 1$ if $j \le R_k$ and $r_{k,j} = 0$ otherwise. Then

$$\phi(R_k) = \phi(0) + \sum_{j=1}^{M} c_j r_{k,j} \tag{9.4}$$

and the optimization problem can be recast as in Fig. 9.11. Notice that condition (9.5c) enforces the fact that, if $r_{k,j} = 1$ then $r_{k,j'} = 1$ for all $1 \le j' \le j$, in other words, if data segment x_j is included in the first k data layers, then all previous segments are also part of the first k data layers.

9.4.2   Network code construction

In this section, we present a polynomial time algorithm which constructs a linear network code for the given network, such that the optimized flow rates for all the sinks are achieved.

9.4.2.1   Algorithm description

Given the optimized rates $R_1, \ldots, R_N$, and the flow in each multicast layer, we want to construct a layered multicast code which achieves the rate R_k at each sink in T_k, $1 \le k \le N$.

We choose the maximum rate R_N as the message dimension, i.e., the source transmits R_N source segments (or messages) in a unit time. Our framework guarantees that any sink in subset T_k receives R_k messages, which are the linear combinations of the first R_k

source segments. Moreover, these messages are linearly independent, thus ensuring the decodability of all first R_k source segments.

Note that any network G can be converted to an equivalent unit-capacity network G'. Moreover, the solution to the data flow optimization problem for network G can be easily converted to an equivalent solution for network G'. Therefore we assume without restricting the generality that all edges of G have unit capacity.

Based on the values $x_{i,j}^{t,k}$ output by the flow optimization algorithm, we construct for each sink $t \in T$ a set of $R_{L(t)}$ edge-disjoint paths $Q_t^1, \ldots, Q_t^{R_{L(t)}}$ from s to t. Note first that due to condition (9.2k) and the fact that any edge has capacity 1, it follows that any edge may carry flow in a single layer. If edge e carries flow in layer l, i.e., $y_e^l = 1$, we say that edge e is in layer l and denote $L(e) = l$. The edge-disjoint paths $Q_t^1, \ldots, Q_t^{R_{L(t)}}$ are constructed such that the following conditions must be satisfied.

(C1) All the paths contain only edges in layers 1 through $L(t)$.

(C2) For any path Q and any two consecutive edges e_1 and e_2 of Q, edge e_1 is in a lower or the same layer as e_2, i.e., $L(e_1) \leq L(e_2)$.

(C3) For any $i, 1 \leq i \leq R_{L(t)} - 1$, the first edge of Q_t^i is in a lower or the same layer as the first edge in Q_t^{i+1}.

The existence of such paths satisfying the above requirements is ensured by the constraints imposed on the node potentials (9.2d–9.2i). Further, for an edge e in a path from s to $t \in T$, let $\phi_t(e)$ denote the predecessor edge on the path and $T(e)$ denote the set of sinks using e in the flow paths. Then, Algorithm 2, which is inspired by the LIF algorithm [48], constructs a linear network code such that the optimized flow rates are achieved.

The algorithm constructs an R_N-dimensional global encoding vector over a finite field F with $|F| > |T|, f(e) = (f_1(e), f_2(e), \ldots f_{R_N}(e))$, for each edge e which carries flow to some sink.

The key idea in order to ensure the algorithm correctness is to maintain an invariant that for each sink t there is a set C_t of $R_{L(t)}$ edges such that the global encoding vectors in the set $\{f(c) : c \in C_t\}$ are linearly independent and, moreover, $f_j(c) = 0$ for all $j, R_{L(c)} + 1 \leq j \leq R_N$. The meaning of the latter condition is that since edge c is in layer $L(c)$, the message passed along this edge can only be a function of source segments $x_1, \ldots, x_{R_{L(c)}}$. Furthermore, the set C_t must contain an edge from each path Q_t^i, $1 \leq i \leq R_{L(t)}$, and at the end of the algorithm we must have $C_t \subseteq In(t)$.

9.4.2.2 Proof of correctness

The correctness of Algorithm 2 follows from the following lemmas.

LEMMA 9.2 *Assign each imaginary edge e_t^i to a layer as follows. Let $L\left(e_t^i\right) = k$ if and only if $R_{k-1} < i \leq R_k$. Then, after assigning the imaginary edges to the $s - t$ paths, condition (C2) is still satisfied for all flow paths. Moreover, the invariant holds at the initialization step.*

for *each $e \in E$* **do**
  | Set $f(e) = [0^{R_N}]$;
end
for *each $t \in T$* **do**
  | Construct $R_{L(t)}$ edge-disjoint paths $\left\{Q_t^1, \ldots, Q_t^{R_{L(t)}}\right\}$ from s to t such that
  | conditions C1–C3 are satisfied;
end
Insert a super source s' into V
for *each $t \in T$* **do**
  | Add $R_{L(t)}$ parallel imaginary edges $\left\{e_t^1, \ldots, e_t^{R_{L(t)}}\right\}$ from s' to s into E;
  | Set $f\left(e_t^i\right) = [0^{i-1}, 1, 0^{R_N - i}]$;
  | Assign e_t^i to a path Q_t^i;
  | Set $C_t = \left\{e_t^1, \ldots, e_t^{R_{L(t)}}\right\}$;   —(*)
end
for *each node $t' \in V \backslash \{s'\}$ in topological order* **do**
  | **for** *each edge $e \in Out(t')$* **do**
  | Choose a global coding vector $f(e)$ such that $f_j(e) = 0$ for all
  | $j, R_{L(e)} + 1 \leq j \leq R_N$, and
  | $\forall t \in T(e), f(e)$ is linearly independent of $\{f(c) : c \in C_t \setminus \{\phi_t(e)\}\}$;
  | —(**)
  | **for** *each $t \in T(e)$* **do**
  | | $C_t = (C_t \backslash \{\phi_t(e)\}) \cup \{e\}$;
  | **end**
  | **end**
end

Algorithm 2: (Construction of intra-layer linear network code) the objective is to construct an R_N-dimensional F-valued linear network code achieving the rate $R_{L(t)}$ for each sink node $t \in T$, when $|F| > |T|$.

Proof The fact that the invariant holds at the initialization step is obvious. It remains to prove the first claim. For any sink t, and any $k, 1 \leq k \leq R_{L(t)}$, let $n(t, k)$ denote the number of $s - t$ paths for which the first edge (before the inclusion of imaginary edges) is in layer k. According to the source potential constraints (9.2d) and (9.2e), we have

$$\sum_{i=1}^{j} n(t, i) \leq R_j, \quad 1 \leq j < L(t) \tag{9.6a}$$

$$\sum_{i=1}^{j} n(t, i) = R_j, \quad j = L(t). \tag{9.6b}$$

The above conditions together with (C3) imply that the first edge in the path Q_t^i (before the inclusion of the imaginary edge e_t^i in the path) is in at least k-th layer, where $R_{k-1} < n \leq R_k$. Since $L(e_t^i) = k$ the conclusion of the lemma follows. $\qquad \square$

LEMMA 9.3   *The global coding kernel $f(e)$ in step $(**)$ can be found, when* $|F| > |T|$.

Proof   This proof closely follows the proof of Lemma 4 in [48]. Let $P(e) = \{\phi_t(e) : t \in T(e)\}$ denote the set of predecessor edges of e in some flow paths. The global encoding vector $f(e)$ is constructed by first finding a local encoding vector $(k_e(e') : e' \in P(e))$ and setting

$$f(e) = \sum_{e' \in P(e)} k_e(e') f(e'). \tag{9.7}$$

Since all flow paths satisfy conditions (C2) it follows that $L(e') \leq L(e)$ for all $e' \in P(e)$. By the invariant, we have $f_j(e') = 0$ for all $j, R_{L(e')} + 1 \leq j \leq R_N$. Hence, (9.7) will further ensure that $f_j(e) = 0$ for all $j, R_{L(e)} + 1 \leq j \leq R_N$.

It remains to show that there exists a local encoding vector $(k_e(e') : e' \in P(e))$ such that $f(e)$ is linearly independent of $\{f(c) : c \in C_t \setminus \{\phi_t(e)\}\}$ for any $t \in T(e)$. By condition (C1), we have $L(c) \leq L(t)$ for all $c \in C_t$; hence the last $R_N - R_{L(t)}$ components of $f(c)$ are zeros. Then, for each $t \in T(e)$ and $c \in C_t$, let $f'(c)$ denote the $R_{L(t)}$-dimensional vector obtained from $f(c)$ after removing the last $R_N - R_{L(t)}$ components. Define $f'(e)$ in the same manner. Clearly, relation (9.7) still holds if f is replaced by f', i.e.,

$$f'(e) = \sum_{e' \in P(e)} k_e(e') f'(e'). \tag{9.8}$$

Note that due to the invariant, the set of vectors $\{f'(c) : c \in C_t\}$ forms a basis of $F^{R_{L(t)}}$. Then, when writing $f'(e)$ as a linear combination of the vectors in this basis, the coefficient assigned to basis vector $f'(\phi_t(e))$ must be $k_e(\phi_t(e)) + \alpha$ for some uniquely determined α which does not depend on $k_e(\phi_t(e))$.

It follows that for any choice of $\{k_e(e') : e' \in P(e) \setminus \{\phi_t(e)\}\}$, there is one and only one $k_e(\phi_t(e))$ to make $f(e)$ linearly dependent of $\{f(c) : c \in C_t \setminus \{\phi_t(e)\}\}$, namely, $k_e(\phi_t(e)) = -\alpha$. So there are $|F|^{|P(e)|-1}$ invalid local coding vectors for a receiver $t \in T(e)$, and the total number of invalid local coding vectors is $N \leq |T| \cdot |F|^{|P(e)|-1} < |F|^{|P(e)|}$. Therefore, there must exist at least one valid local coding vector.   □

Remark 9.4   The new algorithm does not require a larger field size compared to the previous layered multicast scheme.

LEMMA 9.4   *Any sink $t \in T$, is guaranteed to receive $R_{L(t)}$ messages which are linear combinations of the source messages $x_1, \ldots, x_{R_{L(t)}}$. Moreover, these $R_{L(t)}$ messages are linearly independent, thus ensuring the recovering of the first $R_{L(t)}$ data messages.*

Proof   By Lemmas 9.2 and 9.3 the invariant holds at the end of the algorithm. Hence t receives $R_{L(t)}$ messages, carried along the edges in C_t. Furthermore, due to condition (C1), all edges in C_t are in layer $L(t)$ or lower layers. Therefore, these messages are necessarily linear combinations of the data in the first $L(t)$ layers, $x_1, \ldots, x_{R_{L(t)}}$. They are also linear independent by the invariant. Thus the proof is complete.   □

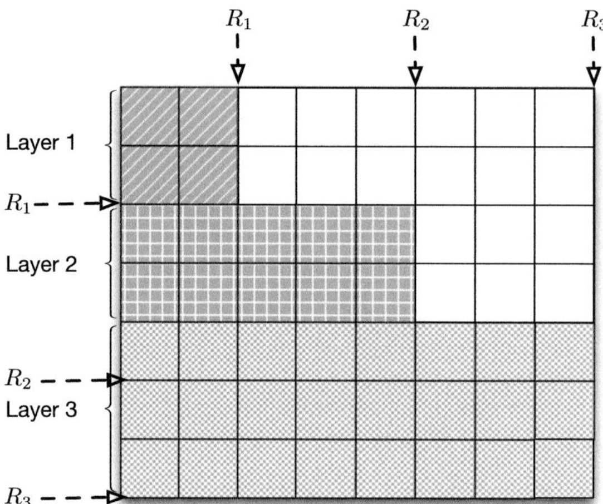

Figure 9.12 Example of data allocation among layers for a sink in layer 3. © 2009 IEEE. Reprinted with permission from [143]

Remark 9.5   For some sink t it is not guaranteed that the messages received over the edges in some multicast layer k are decodable. Moreover, the decodability is not guaranteed either for all messages received over the first k multicast layers, but the decodability of messages received over the first $L(t)$ multicast layers is ensured, and this is all that matters. This concept is illustrated in Fig. 9.12. The example in Fig. 9.12 shows the source segment allocation for some sink t in T_3. The vertical axis denotes the flow received by the sink, while the horizontal axis denotes the source segments. Each row of the matrix can be considered as the global coding vector of a message received at the sink. The shadowed blocks indicate non-zero coefficients and blank blocks indicate zero coefficients. Note that, given the flow over the first 2 layers, sink t cannot decode the first R_2 source segments. However, it can decode the first R_3 source segments received over the first three layers because the sink is guaranteed to receive R_3 units of flow.

9.4.2.3   Complexity analysis

Initializing the imaginary links takes $O\left(R_N^2\right)$ time. Finding a flow augmenting path takes $O(E)$ time. Hence, constructing $R_{L(t)}$ disjoint path for each $t \in T$ takes $O(|E||T|R_N)$ time. The global coding vector $f(e)$ can be found in $O(|T|^2 R_N)$ time, similar to the deterministic implementation in LIF. Combining all the parts, the total running time of Algorithm 1 is $O(|E||T|^2 R_N)$, which is the same as in the previous intra-layer network coding schemes.

9.4.3   Performance evaluation

This subsection contains some simulation results of the inter-layer network coding scheme studied in the section. In the simulations, we consider a family of networks which are first introduced in [48]. In this network model, all the sinks are connected to

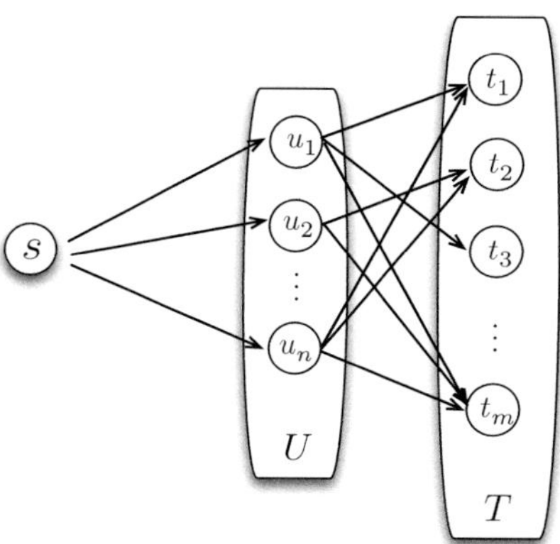

Figure 9.13 Network model used in simulations. © 2009 IEEE. Reprinted with permission from [143]

the source through a group of intermediate nodes (as the network shown in Fig. 9.13). This network model mimics the practical multimedia distribution system with several distributed servers. All of the distributed servers in U connect to a central server s, and each client in T connects to several distributed servers.

The networks used in simulations are randomly generated as follows. We start with the source node s and add intermediate nodes and sink nodes sequentially. We fix the number of intermediate nodes to be 10, and each intermediate node connects directly to the source s. The total number of sinks is N_T, and each sink randomly connects to 50% of the intermediate nodes. Once the network is constructed, we assign a random capacity between 0 to C_1 (kbits/s) to the edges between s and U, and assign a random integer capacity between 0 to C_2 (kbits/s) to the edges between U and T.

We compare the performance of the proposed layered multicast scheme with intra-layer network coding versus the layered multicast with intra-layer network coding of [50] and a simple layered multicast without network coding. We also test the impact of refining the sink partition $\mathcal{T}$, therefore we consider two cases for the proposed scheme: (1) $\mathcal{T} = \mathcal{T}_{\text{max-flow}}$; (2) $\mathcal{T}$ is a refinement of $\mathcal{T}_{\text{max-flow}}$ where each subset contains only one sink. We refer to the above two cases as scheme A and B, respectively. According to the observations in Subsection 9.4.1.4, we expect that scheme B achieves higher performance than scheme A, and that both should outperform the other two methods.

9.4.3.1 Performance comparison of rate-maximized layered multicast schemes

We first compare the performance of all candidate schemes using the rate as fidelity function in the flow optimization problem. In other words, we compare the solutions which maximize the overall received flow. The comparison is with respect to a performance measure called Average Normalized Rate (ANR), which is defined as the ratio of total rate received by all sinks divided by the sum of the max-flow values of all sinks. Clearly, the larger the ANR, the better the scheme. Although the optimal ANR value for

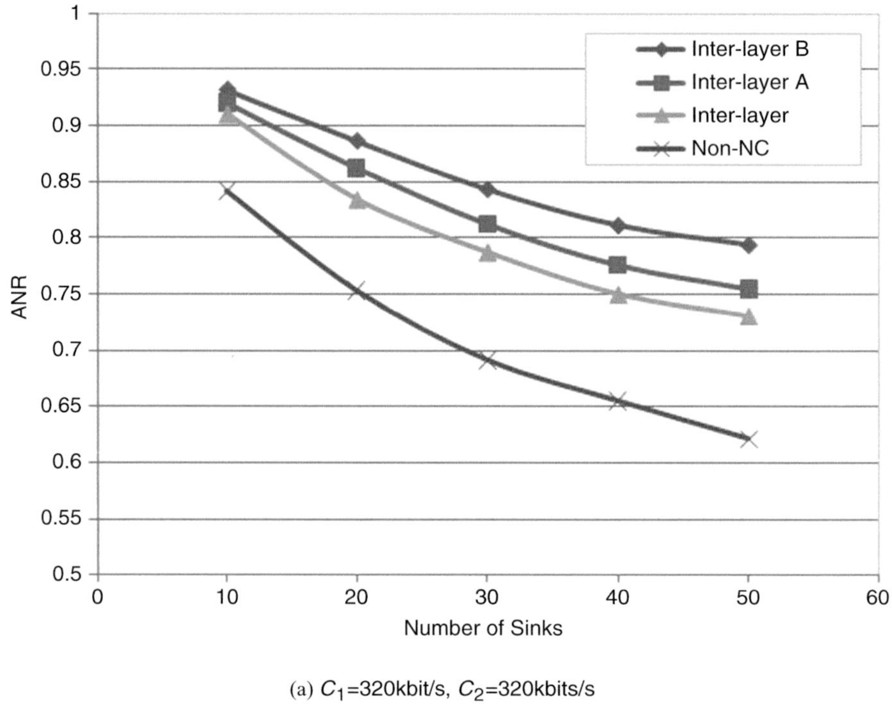

(a) C_1=320kbit/s, C_2=320kbits/s

Figure 9.14 Average normalized rates of four different layered multicast schemes

a certain network is generally unknown, an obvious upper bound of ANR is 1. Since there does not necessary exist a network code that achieves the individual max-flow of all the sinks, the upper bound 1 is not tight for all the networks, even for the optimal solution of multi-rate multicast.

The performance of the four schemes is evaluated for different network sizes, and the results are plotted in Fig. 9.14. We can see from the figure that the proposed intra-layer technique always outperforms the other two opponents. Moreover, scheme B outperforms scheme A, as expected. As the network size increases, the gap to the upper bound of 1 increases for all the schemes. We believe that is due to the fact that as the number of sinks increases, intuitively, the upper bound of 1 becomes looser since it is more difficult to satisfy the max-flow value for all sinks.

The performance comparison also shows the advantage of network coding based multicast schemes over the schemes.

9.4.3.2 Performance comparison of PSNR-maximized layered multicast schemes

We use JSVM 9.15 codec [53] to generate the H.264 SVC [54] confined scalable video streams. By enabling the median grain scalability (MGS) feature in JSVM, we can get a scalable video stream with fine-quality scalability. The rate-PSNR curve of the "foreman" sequence (CIF) is shown in Fig. 9.15, and we use this curve in the following experiments.

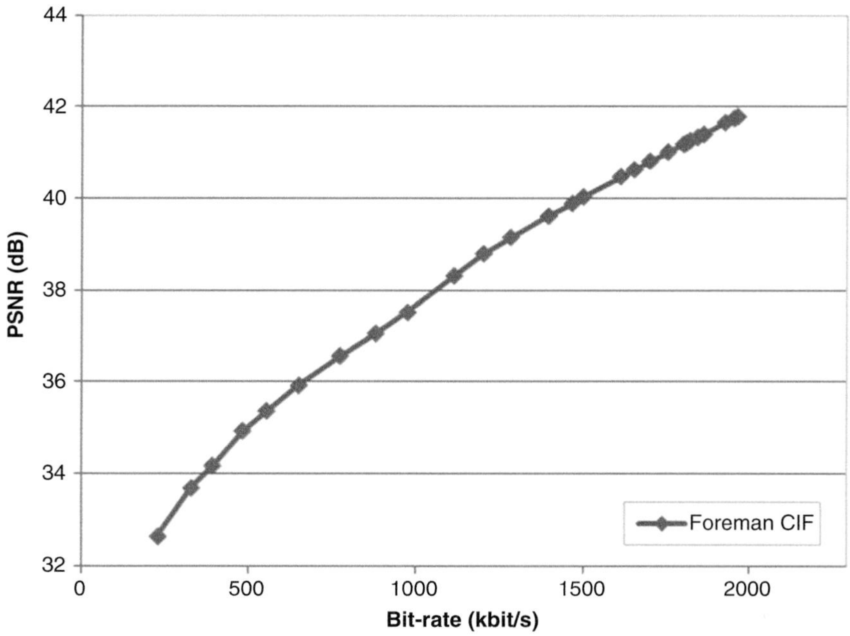

Figure 9.15 Rate-PSNR curve used in the experiments

We first compare the rate-maximized solution (i.e., where $\phi(R) = R$) with the PSNR-maximized solution (where the fidelity function is PSNR) of the proposed layered multicast scheme. In both cases the one sink per subset partition is used (i.e., scheme B). The performance measure is the average PSNR at the sink nodes. Figure 9.16 plots the average PSNR for the PSNR-maximized solution and for the rate-maximized solution. The comparison results in Fig. 9.16 show that maximizing PSNR directly always outperforms maximizing the rate in terms of reconstruction fidelity.

Next, we compare the PSNR-maximized solutions of all candidate schemes. From the results shown in Fig. 9.17, we find that all candidate schemes perform in a similar way as in the simulation using linear cost function. The proposed inter-layer schemes always outperform the intra-layer scheme, and network coding based schemes greatly outperform non–network coding based schemes. As the network size increases, the achieved average PSNR of all schemes decreases. However, the intra-layer scheme B has the lowest decrease rate.

9.4.4 Conclusions

In this chapter, we studied the idea of multicasting progressive codes using network coding. With the aid of layered coding, receiver nodes with higher bandwidths are able to receive more encoding layers and hence reconstruct the source at a higher quality. Network coding will allow for efficient communication of these encoding layers. We

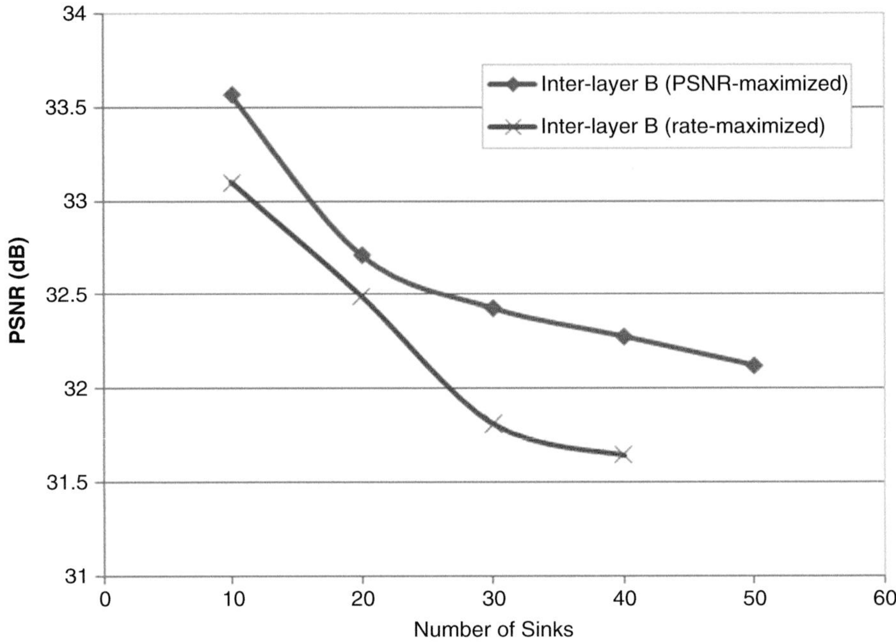

Figure 9.16 PSNR comparison between solutions optimizing PSNR directly and optimizing flow

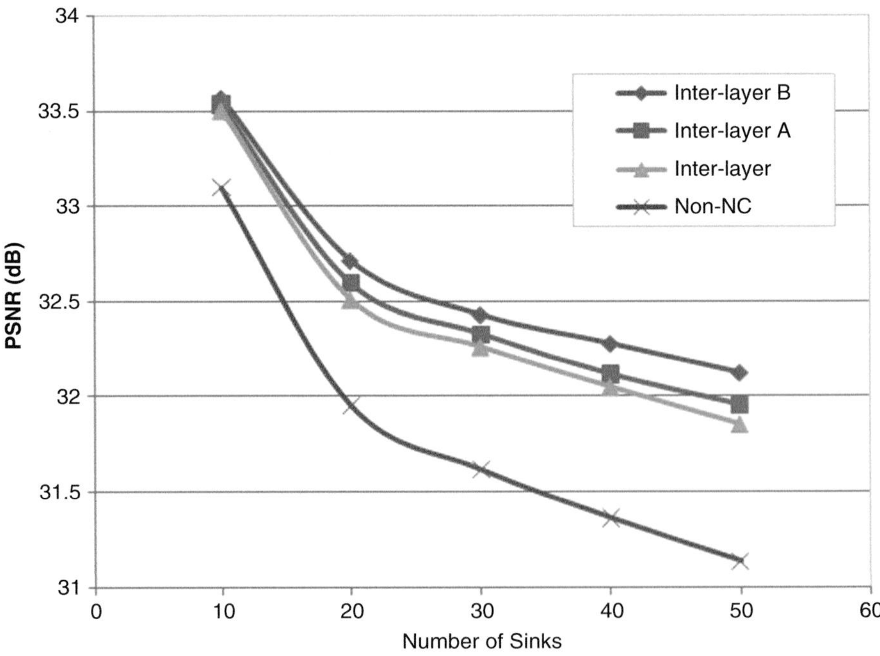

Figure 9.17 Average PSNR of different layered multicast schemes

studied scenarios in which this method is guaranteed to be absolutely optimal, i.e., each receiver can reconstruct the source at the highest quality, as allowed by its individual max-flow. For other scenarios, we studied algorithms for optimization of the rate of the encoding layers and their network information flow such that the average reconstruction quality at all receivers is maximized. So far, we only considered cases with a single source. The next chapter will deal with the case where multiple, perhaps correlated sources have to communicate simultaneously.

10 Lossy communication of multiple correlated sources

So far, we have considered lossy multicast problems involving a single source, with extensions to applications with multi-uncorrelated sources. In this chapter, we investigate the problem of communicating multiple correlated sources, simultaneously, to a set of intended receivers. Most of the results known in this category are for the cases with at most two sources and receivers. Unfortunately, very little is known for more general cases. Even for the case of two sources and sink nodes, many variations of the problem are still open. In the rest of this chapter, we will review the known results in detail.

10.1 Simple Wyner-Ziv

The Wyner-Ziv problem deals with the coding of a source X with a side information Y, which is correlated with X, known only at the decoder (Fig. 10.1). This problem is different from the traditional problem of conditional lossy coding in which side information is known at both the encoder and decoder.

Let X and Y be discrete random variables taking values in alphabets $\mathcal{X}$ and $\mathcal{Y}$. The joint distribution of X and Y is $p(x, y)$. Let $(X_1, Y_1), (X_2, Y_2), \ldots, (X_n, Y_n)$ be a sequence of i.i.d. random variables generated by (X, Y). A distributed source code is composed of one encoder

$$f : \mathcal{X}^n \to \{1, 2, ..., M\},$$

and a decoder

$$g : \{1, 2, ..., M\} \times \mathcal{Y}^n \to \hat{\mathcal{X}}^n.$$

The distortion is defined as

$$d(x, \hat{x}) = \frac{1}{n} \sum_{i=1}^{n} d(x, \hat{x}_i),$$

where d is a distortion measure $d : \mathcal{X} \times \hat{\mathcal{X}} \to R_+$.

The rate-distortion function with side information $R_Y^{WZ}(D)$ is defined as the infimum of rate R such that there exist encoder and decoder such that for arbitrary $\epsilon > 0$, and sufficiently large n,

$$\frac{1}{n} logM \leq R + \epsilon,$$

$$Ed(X^n, g(f(X^n), Y^n)) \leq D + \epsilon.$$

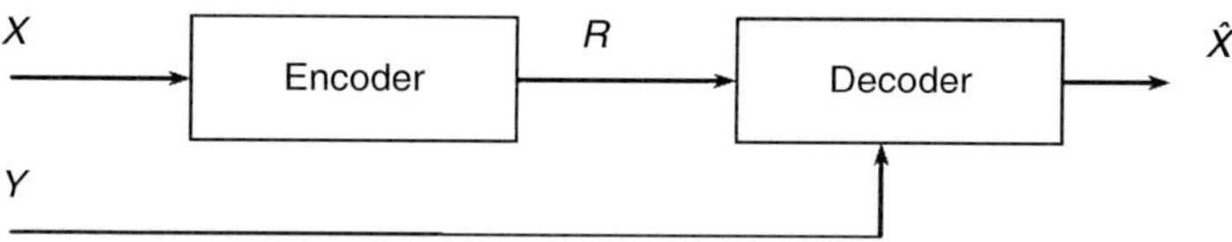

Figure 10.1 The Wyner-Ziv problem (source coding with side information known only at the decoder)

10.2 The Wyner-Ziv theorem and its proof

THEOREM 10.1 (Wyner-Ziv)   *Given $p(x,y)$, the Wyner-Ziv rate-distortion function with side information is*

$$R_Y^{WZ}(D) = \min_{Z,f}(I(X;Z) - I(Y;Z)),$$

where Z is an auxiliary random variable jointly distributed with X and Y, taking values in a discrete alphabet $\mathcal{Z}$ satisfying $|\mathcal{Z}| \le |\mathcal{X}| + 1$, and $Y - X - Z$ forms a Markov chain. And the minimization is over all possible Z and function f, such that

$$Ed(X,f(Y,Z)) \le D.$$

Before proving this theorem, we will introduce the concept of strong typicality and Markov lemma.

10.2.1 Strong typicality and Markov lemma

n-sequence $x \in \mathcal{X}^n$ is said to be ϵ-strongly typical with respect to a distribution $p(a)$ on $\mathcal{X}$, denoted by $x \in T_{[p],\epsilon}^n$, if

$$\left|\frac{1}{n}N(a|x) - p(a)\right| < \frac{\epsilon}{|\mathcal{X}|}, \text{ for every } a \in \mathcal{X},$$

and no $a \in \mathcal{X}$ occurs in x with $p(a) = 0$. $N(a|x)$ is the number of occurrence of a in x. Similarly, n-sequences $x \in \mathcal{X}^n$ and $y \in \mathcal{Y}^n$ are said to be ϵ-strongly typical with respect to a joint distribution $p(a,b)$ on $\mathcal{X} \times \mathcal{Y}$, denoted by $(x,y) \in T_\epsilon^n(X,Y)$, if

$$\left|\frac{1}{n}N(a,b|x,y) - p(a,b)\right| < \frac{\epsilon}{|\mathcal{X}||\mathcal{Y}|}, \text{ for every } (a,b) \in \mathcal{X} \times \mathcal{Y},$$

and no $(a,b) \in \mathcal{X} \times \mathcal{Y}$ occurs in (x,y) with $p(a,b) = 0$. $N(a,b|x,y)$ is the number of occurrence of pair (a,b) in (x,y). Strong typicality has a close relationship with the method of types. The type of a sequence $x \in \mathcal{X}^n$ is the distribution P_x on $\mathcal{X}$ defined by

$$P_x(a) = \frac{1}{n}N(a|x), \forall a \in \mathcal{X}.$$

The joint type of a pair of n-sequences $x \in \mathcal{X}^n$ and $y \in \mathcal{Y}^n$ can be similarly defined. It is clear that strongly typical sequences are those sequences whose type is very close to the given distribution function. Detailed results of method of types can be found in [110] and [109]. The typical set $T_\epsilon^n(X, Y)$ is briefly denoted by $T(X, Y)$ according to the delta-convention [110]. The most important property of strong typicality is the following.

(*Property of strong typicality*): There exists a sequence $\epsilon_n \to 0$ depending only on $|\mathcal{X}|$ so that for every distribution P on $\mathcal{X}$

$$Pr\left(T_{[P]}^n\right) \geq 1 - \epsilon_n,$$

$$\left|\frac{1}{n}\log\left|T_{[P]}^n\right| - H(P)\right| \leq \epsilon_n$$

where $H(P)$ is the entropy of the distribution.

The above properties mean that the cardinality of the set of typical sequences is almost $2^{nH(P)}$, while its probability approaches 1 as $n \to \infty$.

It is true that if a triple of sequences x, y, and z are strongly jointly typical, any two of them are also strongly jointly typical. But the converse is not true, $(x, y) \in T(X, Y)$ and $(y, z) \in T(Y, Z)$ do not guarantee the joint typicality of x, y and z. But this becomes asymptotically true if $X - Y - Z$ forms a Markov chain in this order. We summarize this in the following lemma.[1]

LEMMA 10.1 (Markov Lemma)　*Let X, Y, Z form a Markov chain $X - Y - Z$. $X = (X_1, X_2, ..., X_n)$ and $Y = (Y_1, Y_2, ..., Y_n)$ are i.i.d. sequences drawn according to $p_{XY}(x, y)$. Then for any $\epsilon > 0$*

$$\lim_{n \to \infty} Pr\left[(X, z) \in T_\epsilon^n(X, Z) | (Y, z) \in T_\epsilon^n(Y, Z)\right] = 1.$$

10.2.2　Proof of the Wyner-Ziv theorem

Achievable part:
Given random variable Z satisfying $Z - X - Y$ and a function $f(Z, Y)$, the procedure is shown as follows.

Random code book generation:
Generate $2^{nI(X;Z)}$ codewords by arbitrarily picking up typical sequences in $T(Z)$. Then randomly put these codewords into $2^{n(I(X;Z)-I(Y;Z))}$ bins with approximately $2^{nI(Y;Z)}$ codewords in each bin. And reveal this assignment to both encoder and decoder.

Encoding:
Given a source sequence x, the encoder looks for a codeword z such that (x, z) is jointly typical. The encoder then sends the index of the bin in which z belongs. The code rate is thus approximately $I(Z; X) - I(Y; Z)$. If no such sequence z is found, send an arbitrary index and an error is declared.

Decoding:
On receiving the index of the bin, say i, the decoder searches in this bin and tries to find a sequence y such that (z, y) is jointly typical. Then the decoder uses the function f to

---

[1]　Here we introduce the form of Berger, which requires the independent and identical distribution of (X_n, Y_n). There are also generalized forms of Markov lemma without this assumption.

generate the reconstruction $\hat{x} = f(y, z)$. If no such sequence y is found, the decoder uses an arbitrary sequence y to calculate the reconstruction.

Analysis of probability of error:

The encoding-decoding procedure fails if and only if at least one of the following events occurs.

E_0: x and y are not jointly typical;

The probability that x and y are not jointly typical approaches one according to the property of strong joint typicality.

E_1: There is no codeword z such that (x, z) is jointly typical.

For any typical sequence x, the probability that any sequence in $T(Z)$ is strongly jointly typical with x is approximately

$$\frac{2^{nH(Z|X)}}{2^{nH(Z)}} = 2^{-nI(X;Z)}.$$

So, if the size of the codebook is greater than $2^{nI(X;Z)}$, the probability of E_1 approaches zero as n approaches infinity.

E_2: (x, z) is jointly typical but (y, z) isn't.

The probability of E2 approaches to zero according to the Markov lemma.

E_3: There exists another sequence z' in the same bin as z such that (y, z') is jointly typical.

The probability of an arbitrarily chosen z is jointly typical with y is $2^{-nI(Y;Z)}$, so, if the size of each bin is smaller than $2^{nI(Y;Z)}$, the probability of E_3 approaches zero.

Converse part:

Consider any Wyner-Ziv code with encoder f_n and decoder g_n and satisfying the distortion constraint. We need to prove that there exist a random variable Z and function h, such that $R \geq I(X; Z|Y)$ and $Ed(X, h(Y, Z)) \leq D$. Denote $S = f(X^n)$, $X^{i-1} = (X_1, ..., X_{i-1})$, $Y_{i+1}^n = (Y_{i+1}, ..., Y_n)$, and $Z_i = \left(S, Y^{i-1}, Y_{i+1}^n\right)$. Let I be a random variable independent of all others and uniformly distributed in $N = \{1, ..., n\}$ and $Z = (Z_I, I)$.

$$nR \geq H(S)$$
$$= H(S|Y^n)$$
$$= I(X^n; S|Y^n)$$
$$= H(X^n|Y^n) - H(X^n|SY^n)$$
$$= \sum_{i=1}^{n} H(X_i|Y_i) - \sum_{i=1}^{n} H(X_i|X^{i-1}SY^n)$$
$$\geq \sum_{i=1}^{n} H(X_i|Y_i) - \sum_{i=1}^{n} H(X_i|Z_iY_i)$$
$$= \sum_{i=1}^{n} I(X_i; Z_i|Y_i)$$
$$= nI(X; Z|Y).$$

Denote $g_i(f(X^n), Y^n)$ as the output of the decoder at time i, $1 \leq i \leq n$. Because Z contains $f(X^n)$ and $Y^{n \backslash I}$, define

$$h(Y, Z) = g_I(f(X^n), Y^n).$$

Then according to the assumption for the distortion, we have

$$Ed(X, h(Y, Z)) = \frac{1}{n} \sum_{i=1}^{n} Ed(X, g_i(f(X^n), Y^n)) \leq D.$$

And we also have the Markov chain $Z - X - Y$ according to the definition of Z. The converse part is proved.

Finally, the constraint of the cardinality of Z can be proved using the support lemma [110].

10.3   Wyner-Ziv function for Gaussian sources and binary sources

10.3.1   Gaussian sources

Let's consider the case that X and Y are jointly Gaussian and the distortion criterion is the mean square distortion. The most interesting thing for this quadratic Gaussian case is that it does not have any performance loss comparing with that if the encoder knows the side information, i.e., $R_Y^{WZ}(D) = R_Y(D)$. Without loss of generality, assume X and Y have zero means and unit variances. Let

$$Y = rX + N,$$

where N is a Gaussian random variable with zero mean and variance $1 - r^2$. And N is independent of X.

We have the following result:

$$R_Y^{WZ}(D) = \frac{1}{2} \log \frac{1 - r^2}{D}.$$

This result is an immediate consequence of Theorem 10.1 by letting $Z = X + M$, where M is a Gaussian random variable with zero mean and variance D.

The jointly Gaussian condition can be relaxed. In fact, if there is a virtual AWGN channel connecting X and Y, the Wyner-Ziv solution is the same as above.

10.3.2   Binary sources

Wyner and Ziv [137] showed that if X and Y are doubly binary symmetric sources, where $Y = X + Z$ with Z Bernoulli-p, and the distortion is the Hamming distance, then the Wyner-Ziv rate-distortion function is given by

$$R_Y^{WZ}(D) = l.c.c\{H(p * D) - H(D), (p, 0)\}, 0 \leq D \leq p,$$

where $l.c.c$ denotes lower convex envelop of the function $g(D) = H(p * D) - H(D)$ and the point $(D = p, R = 0)$. $p * D = p(1 - D) + (1 - p)D$ is the binary convolution operation of two real values p and D. The rate-distortion curve is shown in Fig. 10.2. Unlike the

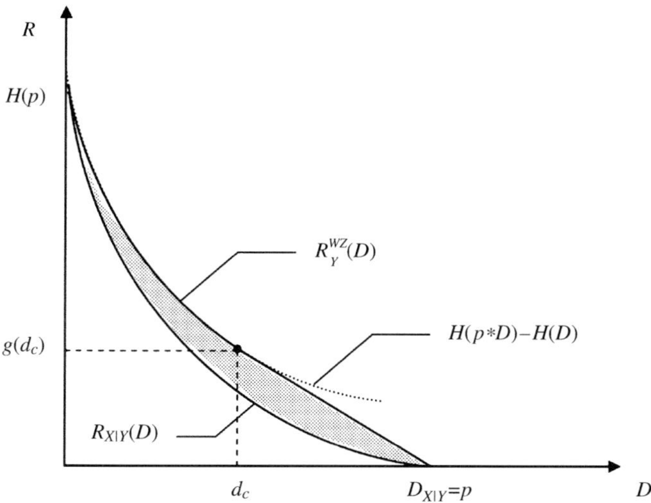

Figure 10.2 The solution of the Wyner-Ziv problem of doubly symmetric binary sources

quadratic Gaussian case, the Wyner-Ziv problem of binary sources is strictly greater than the conditional rate-distortion function $R_{X|Y}(D) = H(p) - H(D)$, corresponding to the case that both encoder and decoder know Y. In Fig. 10.2, the shaded region is the rate loss of the Wyner-Ziv coding comparing with the conditional rate-distortion function. The coordinate of the connection point of $g(D)$ and the straight line is determined by

$$\frac{g(d_c)}{d_c - p} = g'(d_c).$$

10.4   Wyner-Ziv code design

In [139], Zamir *et al.* propose the use of nested codes to construct Wyner-Ziv code. In their setting, a nested linear/lattice code composes of a fine code and a coarse code. The fine code performs as a source code and the coarse code performs as a channel code. For example, the parity check matrices of a nested linear code pair $(\mathcal{C}_1, \mathcal{C}_2)$ satisfying $\mathcal{C}_2 \subset \mathcal{C}_1$ can always be expressed as

$$\mathbf{H}_2 = \begin{bmatrix} \mathbf{H}_1 \\ \Delta \mathbf{H} \end{bmatrix}$$

where $\mathbf{H}_1$ is an $(n - k_1) \times n$ matrix and $\mathbf{H}_2$ is an $(n - k_2) \times n$ matrix, $k_1 > k_2$. That is to say, all codeword in $\mathcal{C}_2$ also belongs to $\mathcal{C}_1$. In the quadratic Gaussian case, it is shown that the Wyner-Ziv function can be achieved asymptotically using nested lattice code as the dimension goes to infinity.

For practical Wyner-Ziv code design, we usually use a nested quantizor and then perform the Slepian-Wolf coding. Readers are referred to [113] and the references therein.

10.5   Problems closely related to Wyner-Ziv coding

In literature, there are kinds of very important problems which are extensions of Slepian-Wolf and Wyner-Ziv coding. Among them are the direct and indirect multi-terminal source coding problem. The direct source coding problem, which is the traditional multi-terminal source coding problem, is a lossy extension of the Slepian-Wolf coding problem. It has been proposed for forty years. Its quadratic Gaussian result is only known very recently. The indirect problem, also known as the CEO problem, is a comparative new coding system proposed merely more than a decade ago. These two problems are of central importance in multi-user information theory, and in the following, we will give a brief introduction to both of them.

10.5.1   The (direct) multi-terminal source coding problem

Given sources X and Y determined by joint distribution $p(x,y)$. Let (X^n, Y^n) be i.i.d. sequences generated by (X, Y). The two encoder functions are

$$\phi_1 : \mathcal{X}^n \to \{1, 2, ..., M_1\}$$
$$\phi_2 : \mathcal{Y}^n \to \{1, 2, ..., M_2\},$$

and the rates are constrained by

$$\frac{1}{n} log M_i \le R_i + \epsilon, i = 1, 2.$$

The decoder function $\psi = (\psi_1, \psi_2)$ is

$$\psi_1 : \{1, 2, ..., M_1\} \times \{1, 2, ..., M_2\} \to \mathcal{X}^n$$
$$\psi_2 : \{1, 2, ..., M_1\} \times \{1, 2, ..., M_2\} \to \mathcal{Y}^n$$

and the reconstructions are denoted by $\hat{X}^n$ and $\hat{Y}^n$. The vector distortion $\boldsymbol{D} = (\Delta_1, \Delta_2)$ is defined by

$$\Delta_1 = E\frac{1}{n} \sum_{i=1}^{n} d_1(X_i, \hat{X}_i)$$

$$\Delta_2 = E\frac{1}{n} \sum_{i=1}^{n} d_2(Y_i, \hat{Y}_i)$$

where $d_1 : \mathcal{X} \times \mathcal{X} \to [0, \infty)$ and $d_2 : \mathcal{Y} \times \mathcal{Y} \to [0, \infty)$ are distortion measures.

Given (D_1, D_2), a rate pair (R_1, R_2) is admissible if for any $\epsilon > 0$, there exist encoders and decoder (ϕ_1, ϕ_2, ψ), such that, for sufficiently large n, $\Delta_i \le D_i + \epsilon, i = 1, 2$. Setting $\mathcal{R}(D_1, D_2) = \{(R_1, R_2) : (R_1, R_2)\ is\ admissible\}$, the task of the (direct) multi-terminal source coding is to determine the region $\mathcal{R}(D_1, D_2)$. In the following, we give an inner bound and an outer bound.

(*Berger-Tung inner bound*): Denote $\mathcal{R}_{in}(D_1, D_2)$ to be the set of all rate pairs such that there exist auxiliary random variables U and V for which the following conditions are satisfied.

(i) $U - X - Y - V$ form a Markov chain in this order;

(ii) $R_1 \geq I(X; U|V)$;

(iii) $R_2 \geq I(Y; V|U)$;

(iv) $R_1 + R_2 \geq I(XY; UV)$;

(v) There exist reconstruction functions $\hat{X}(U, V)$ and $\hat{Y}(U, V)$ such that $Ed_1(X, \hat{X}) \leq D_1$ and $Ed_2(Y, \hat{Y}) \leq D_2$.

(*Berger-Tung outer bound*): Denote $\mathcal{R}_{out}(D_1, D_2)$ to be the set of all rate pairs such that there exist auxiliary random variables U and V for which the following conditions are satisfied.

(i) $U - X - Y, X - Y - V$;

(ii) $R_1 \geq I(X; U|V)$;

(iii) $R_2 \geq I(Y; V|U)$;

(iv) $R_1 + R_2 \geq I(XY; UV)$;

(v) There exist reconstruction functions $\hat{X}(U, V)$ and $\hat{Y}(U, V)$ such that $Ed_1(X, \hat{X}) \leq D_1$ and $Ed_2(Y, \hat{Y}) \leq D_2$.

THEOREM 10.2 $\mathcal{R}_{in}(D_1, D_2) \subseteq \mathcal{R}(D_1, D_2) \subseteq \mathcal{R}_{out}(D_1, D_2)$.

The exact achievable region for general sources is still open. The most interesting case studied in literature is the quadratic Gaussian case. Although it seems that it would be easier to find the achievable region for this case (considering that many other problems are open for general sources but have closed-form solutions for quadratic Gaussian), surprisingly, the final solution is known only very recently, nearly 30 years after it was proposed. In the following, we will give some detailed results for the quadratic Gaussian multi-terminal problem. Without loss of generality, the Gaussian source (X, Y) is determined by the covariance matrix

$$\begin{pmatrix} 1 & \rho \\ \rho & 1 \end{pmatrix}.$$

Berger-Tung inner bound for Gaussian sources:

$$\mathcal{R}_{in}(D_1, D_2) = \{(R_1, R_2) : R_1 \geq \frac{1}{2} \log^+ \left[(1 - \rho^2) \left[s_1 - \frac{2\rho^2 s_1 s_2}{(1 - \rho^2)\beta} \right]^{-1} \right],$$

$$R_2 \geq \frac{1}{2} \log^+ \left[(1 - \rho^2) \left[s_2 - \frac{2\rho^2 s_1 s_2}{(1 - \rho^2)\beta} \right]^{-1} \right],$$

$$R_1 + R_2 \geq \frac{1}{2} \log^+ \left[\frac{(1 - \rho^2)\beta}{2 s_1 s_2} \right],$$

$$0 < s_1 \leq D_1, 0 < s_2 \leq D_2\}$$

where $\log^+ x = max(\log x, 0)$, $\beta = 1 + \sqrt{1 + \frac{4\rho^2 s_1 s_2}{(1 - \rho^2)^2}}$.

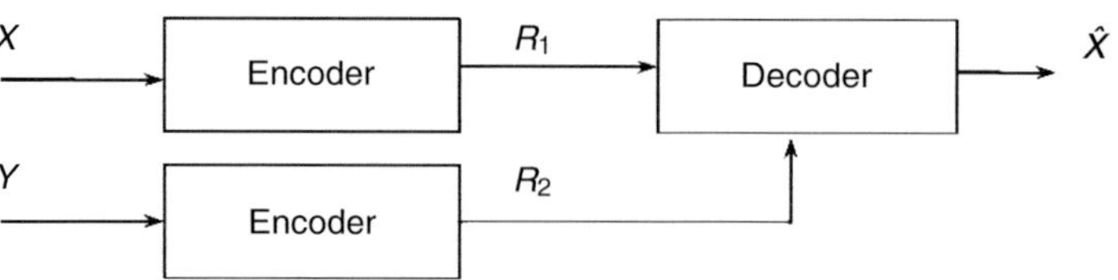

Figure 10.3 Source coding with partial side information

In order to attack the quadratic Gaussian multi-terminal problem, Oohama studied some degraded version of this problem, which is also known as source coding with partial side information for general sources (Fig. 10.3). In this setting, the given distortion level D_2 is sufficiently large. Then the achievable region is defined formally by

$$\mathcal{R}_1(D_1) = \{(R_1, R_2) : (R_1, R_2) \in \mathcal{R}(D_1, D_2) \text{ for some} D_2 > 0\}.$$

$\mathcal{R}_2(D_2)$ is defined similarly. In [114], the exact forms of these functions are determined by

$$\mathcal{R}_1(D_1) = \left\{(R_1, R_2) : R_1 \geq \frac{1}{2} \log^+ \left[\frac{1}{D_1}\left(1 - \rho^2 + \rho^2 2^{-2R_2}\right)\right]\right\}$$

$$\mathcal{R}_2(D_2) = \left\{(R_1, R_2) : R_2 \geq \frac{1}{2} \log^+ \left[\frac{1}{D_2}\left(1 - \rho^2 + \rho^2 2^{-2R_1}\right)\right]\right\}.$$

It is obvious that both $\mathcal{R}_1(D_1)$ and $\mathcal{R}_2(D_2)$ are outer bounds of the problem. Considering that the rate-distortion function of the vector source (X, Y) is a natural outer bound of the sum rate, the outer bound of the quadratic Gaussian multi-terminal problem given by Oohama is

$$\mathcal{R}_{out}(D_1, D_2) = \mathcal{R}_1(D_1) \cap \mathcal{R}_2(D_2) \cap \mathcal{R}_{12}(D_1, D_2),$$

where $\mathcal{R}_{12}(D_1, D_2) = \left\{(R_1, R_2) : R_1 + R_2 \geq \frac{1}{2} \log^+ \left[\frac{1}{D_1 D_2}\left(1 - \rho^2\right)\right]\right\}$, which is the rate-distortion function of the source (X, Y) at the distortion level (D_1, D_2).

It is straightforward to prove that the Berger-Tung inner bound for Gaussian sources is equivalent to the following region:

$$\mathcal{R}_{in}(D_1, D_2) = \mathcal{R}_1(D_1) \cap \mathcal{R}_2(D_2) \cap \tilde{\mathcal{R}}_{12}(D_1, D_2),$$

where

$$\tilde{\mathcal{R}}_{12}(D_1, D_2) = \left\{(R_1, R_2) : R_1 + R_2 \geq \frac{1}{2} \log^+ \left[\frac{(1 - \rho^2)}{2 D_1 D_2}\left(1 + \sqrt{1 + \frac{4\rho^2 D_1 D_2}{(1 - \rho^2)^2}}\right)\right]\right\}.$$

It can be seen that the outer bound and the inner bound partly coincide with each other. Recently, Wanger *et al.* [135] proved that the Berger-Tung inner bound is actually tight for the quadratic Gaussian multi-terminal coding problem. The main idea behind their proof is coupling the multi-terminal problem into a CEO problem (which is also known as an indirect multi-terminal problem and will be discussed later) and thus the techniques of Oohama can be applied.

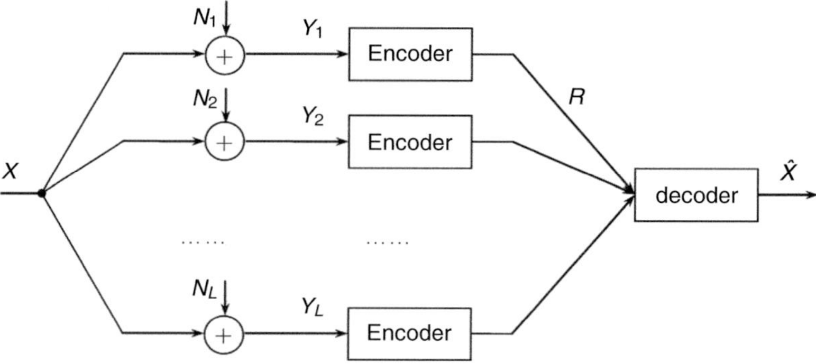

Figure 10.4 The quadratic Gaussian CEO problem. N_i are additive Gaussian noise independent of each other and X, $i = 1, \ldots, L$

THEOREM 10.3 (Rate-distortion region for the Gaussian two-terminal coding):

$$\mathcal{R}(D_1, D_2) = \mathcal{R}_{in}(D_1, D_2) = \mathcal{R}_1(D_1) \cap \mathcal{R}_2(D_2) \cap \tilde{\mathcal{R}}_{12}(D_1, D_2).$$

10.5.2 The CEO problem (indirect multi-terminal source coding)

The CEO problem describes the following coding/transmission scenario. A Chief Executive Officer (CEO) wants to gather information of a remote source sequence which cannot be observed directly. He sends a team of L agents to gather information for him. The agents do not cooperate with each other. And each agent observes independently corrupted versions of the source (see Fig. 10.4). It was shown that if the agents are not allowed to convene, there does not exist a finite value of rate for which even infinitely many agents can make the distortion arbitrarily small [106].

In the above setting, only discrete sources are considered. But of course, analog sources are also very important and people want to find their rate-distortion behavior. The most interesting case is again the Gaussian source with mean square error distortion. In literature, Viswanathan [134] studied this problem from the viewpoint of statistics, while Oohama [115] proved Viswanathan *et al.*'s conjecture using a traditional information theoretical approach. For a Gaussian source X with zero mean and variance σ_X^2, the rate lower bound required for each agent to achieve the distortion D is the following:

$$R(D) = \frac{\sigma_N^2}{2\sigma_X^2} \left(\frac{\sigma_X^2}{D} - 1 \right) + \frac{1}{2} \log \left(\frac{\sigma_X^2}{D} \right).$$

In order to understand the asymptotic behavior of the problem as the coding rate of each individual agent approaches infinity, Viswanathan and Berger define the following function:

$$\beta \left(\sigma_X^2, \sigma_N^2 \right) = \lim_{R \to \infty} R \frac{D(R)}{\sigma_X^2},$$

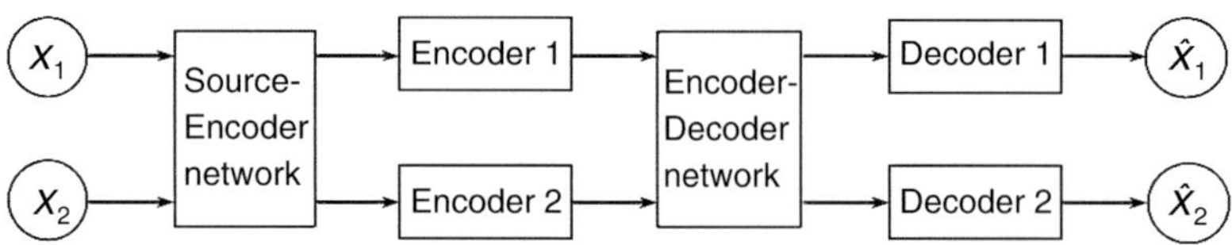

Figure 10.5 Source-encoder-decoder system

where $D(R)$ is the distortion-rate function of the CEO system. It is conjectured by Viswanathan and Berger and finally proved by Oohama that

$$\beta\left(\sigma_X^2, \sigma_N^2\right) = \frac{\sigma_N^2}{2\sigma_X^2}.$$

This means that comparing with cooperated agents, isolated agents suffer from a considerable performance degradation.

The quadratic Gaussian CEO problem can be generalized to the case that independent observations of agents have different variances. In literature, there is also an extension of the CEO problem, the so-called many-help-one problem in the sense that only one of the agents can observe the remote source without noise. The source coding with partial side information is also called one-help-one problem. And the tight achievable region is known as long as any two observations are independent with each other conditioned on the source. It should be noticed that when the noisy channels are not independent, i.e., when the Markov condition no longer holds, the solution of the CEO problem as well as the many-help-one problem are still open. But recently, the Markov condition can be relaxed to some extent and solutions are found for tree structured sources [133].

10.6   A summary of the network source coding problem of no more than two encoders and decoders

A typical source coding problem always includes sources, encoders and decoders. In modern point of view, traditional source coding problem is a special case of lossless/lossy source-network coding. So, when there are multiple encoders and decoders, we usually refer to the encoding-decoding system as a network source coding problem. In this vein, we will pay some attention to the topology structure of two inherent simple networks, i.e., the network between sources and encoder and the network between encoders and decoders. In the following, we give a summary of coding systems of no more than two sources, two encoders and two decoders. The general system is shown in Fig. 10.5.[2]

The source-encoder network and encoder-decoder network are two-in-two-out networks and have no intermediate nodes. Some traditional networks include modulars in Fig. 10.6. In Table 10.1, we summarize the network source coding problems and show whether the corresponding systems have known rate-distortion regions.

---

[2] In [111] [136], more general systems called indirect rate-distortion problems studied. In their system, there is an extra channel between the decoder and the observation.

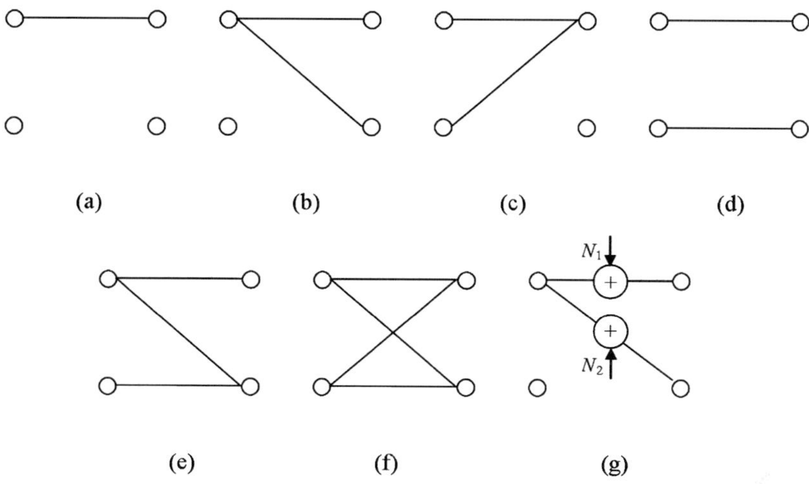

Figure 10.6 Some commonly used source-encoder and encoder-decoder networks in two terminals

When the number of decoders is not restricted to two, there are more variations of the coding systems. For example, [107] deals with the case of multi-terminal source coding with one encoder possibly broken down. They give the whole achievable regions if X_1 is to be reconstructed perfectly in the usual Shannon sense. There is a further generalization

Table 10.1 Network source coding problem of no more than two encoders and decoders

Problem	Source-encoder network	Encoder-decoder network	Reconstruction X_1	X_2	Results
Entropy coding	(a)	(a)	Lossless	/	Closed
Rate-distortion	(a)	(a)	Lossy	/	Closed
Vector R-D	(c)	(b)	Lossy	Lossy	Closed
Successive refinement	(e)	(e)	Lossy	Lossy	Closed
Slepian-Wolf	(d)	(f)	Lossless	Lossless	Closed
Wyner-Ziv	(d)	(e)	Lossless	Lossy	Closed
Partial SI	(d)	(e)	Lossy(/)	Lossy	Open(G)
CEO	(g)	(f)	Lossy	Lossy	Open(G)
Multi-terminal	(d)	(f)	Lossy	Lossy	Open(G)
Multi-terminal with one distortion	(d)	(f)	Lossless	Lossy	Closed

Notes:
SI: Side Information.
Lossy(/) means a lossy reconstruction or not care.
Open(G) means generally open but closed for quadratic Gaussian case.
It should be noted that for multi-terminal source coding with more than two sources, we don't have closed solutions for quadratic Gaussian.

of the problem to the case both encoders may be broken down, which is called multi-terminal with two secondary decoders in [107] and robust distributed source coding in [108]. But the tight achievable region is still unknown even for Gaussian sources. There are also some variations that the input of one encoder is the output of the other encoder. For a systematic analysis of the problem, please refer to [112].

References

[1] D. Slepian and J. K. Wolf. Noiseless coding of correlated information sources. *IEEE Trans. on Inform. Theory*, 19, 1973.

[2] L. H. Ozarow. On a source coding problem with two channels and three receivers. *Bell System Technical Journal*, 59(10), 1980.

[3] T. Ho, M. Medard, M. Effros, and R. Koetter. Network coding for correlated sources. *CISS*, 2004.

[4] A. Ramamoorthy, P. A. Chou, K. Jain, and M. Effros. Separating distributed source coding from network coding. *Allerton Conference on Communication, Control and Computing*, 2004.

[5] http://www.microsoft.com/tv/iptvedition.mspx

[6] http://www.planet.nl/planet/show/id=118880/contentid=668582/sc=780960

[7] http://www.shoutcast.com/ttsl.html

[8] R. Ahlswede, N. Cai, S. Y. R. Li, and R. W. Yeung. Network information flow. *IEEE Trans. on Inform. Theory*, 46, 2000.

[9] S. Y. R. Li, R.W. Yeung, and N. Cai. Linear network coding. *IEEE Trans. on Inform. Theory*, 49, 2003.

[10] A. R. Lehman and E. Lehman. Complexity classification of network information flow problems. *Allerton Communication, and Computing*, 2003.

[11] S. Jaggi, P. Sanders, P. A. Chou, M. Effros, S. Egner, K. Jain, and L. Tolhuizen. Polynomial time algorithms for multicast network code construction. *IEEE Trans. on Inform. Theory*, 51, 2005.

[12] P. Sanders, S. Egner, and L. Tolhuizen. Polynomial time algorithms for network information flow. *ACM SPAA*, 2003.

[13] K. Jain, M. Mahdian, and M. R. Salavatipour. Packing steiner trees. *ACM SODA*, 2003.

[14] S. Hougardy and H. J. Prömel. A 1.598 approximation algorithm for the steiner problem in graphs. In *ACM SODA*, 1999.

[15] G. Robins and A. Zelikovsky. Improved Steiner tree approximation in graphs. *Proceedings of SODA*, 770–779, 2000.

[16] L. C. Lau. An approximate max-steiner-tree-packing min-steiner-cut theorem. In *IEEE FOCS*, 2004.

[17] Nicholas J. A. Harvey, Robert Kleinberg, and April Rasala Lehman. Comparing network coding with multicommodity flow for the k-pairs communication problem. *CSAIL Technical Reports*, 2004.

[18] Nicholas J. A. Harvey, Robert Kleinberg, and April Rasala Lehman. On the capacity of information networks. *IEEE/ACM Trans. Netw.*, 14:2345–2364, 2006.

[19] T. Leighton and S. Rao. Multicommodity max-flow min-cut theorems and their use in designing approximation algorithms. *Journal of the ACM*, 46(6):787–832, 1999.

[20] Z. Li and B. Li. Network coding in undirected networks. *Proc. of CISS*, 2004.

[21] B. Li and Z. Li. Network coding: the case of multiple unicast sessions. *Proceedings of the 42nd Allerton Annual Conference on Communication, Control, and Computing*, 2004.

[22] K. Jain, V. V. Vazirani, and G. Yuval. On the capacity of multiple unicast sessions in undirected graphs. *IEEE/ACM Transactions on Networking*, 14:2805–2809, 2006.

[23] Z. Reznic, R. Zamir, and M. Feder. Joint source-channel coding of a gaussian mixture source over a gaussian broadcast channel. In *IEEE Trans. on Inform. Theory*, 48, 2002.

[24] A. Albanese, J. Blomer, J. Edmonds, M. Sudan, and M. Luby. Priority encoding transmission. *IEEE Trans. on Inform. Theory*, 42, 1996.

[25] R. Venkataramani, G. Kramer, and V. K. Goyal. Multiple description coding with many channels. *IEEE Trans. on Inform. Theory*, 49, 2003.

[26] A. D. Taubman and M. Marcellin. Jpeg2000: Image compression fundamentals, practice and standard. Berlin: Kluwer Academic Publishers, 2001.

[27] W. Li. Overview of fine granularity scalability in mpeg-4 video standard. *IEEE Trans. Circuits Syst. Video Techn.*, 11, 2001.

[28] X. Wu, B. Ma, and N. Sarshar. Rainbow network problems and multiple description coding. *IEEE ISIT*, 2005.

[29] L. Lastras and T. Berger. All sources are nearly successively refinable. *IEEE Trans. Inform. Theory*, 47, 2001.

[30] A. Said and W. Pearlman. A new, fast and efficient image codec based on set partitioning in hierarchical trees. *IEEE Trans. on Circuits and Syst. for Video Tech.*, 6:243–250, 1996.

[31] T. Burger. *Rate Distortion Theory: A Mathematical Basis for Data Compression*. Englewood Cliffs: Prentice-Hall, 1974.

[32] Taekon Kim, Hyun Mun Kim, Ping-Sing Tsai, and T. Acharya. Memory efficient progressive rate-distortion algorithm for jpeg 2000. *IEEE Trans. on Circuits and Systems for Video Tech.*, 15, 2005.

[33] S. Boyd and L. Vandenberghe. Convex Optimization. Online: http://www.stanford.edu/boyd/cvxbook/

[34] S. Dumitrescu, X. Wu, and Z. Wang. Globally optimal uneven error protected packetization of scalable code streams. *IEEE Trans. on Multimedia*, 6, 2004.

[35] R. Puri and K. Ramchandran. Multiple description source coding through forward error correction codes. In *Proc. 33rd Asilomar Conference on Signals, Systems, and Computers*, volume 1, pp. 342–346, 1999.

[36] A. E. Mohr, R. E. Ladner, and E. A. Riskin. Approximately optimal assignment for unequal loss protection. *IEEE ICIP*, 2000.

[37] A. E. Mohr, E. A. Riskin, and R. E. Ladner. Graceful degradation over packet erasure channels through forward error correction. *IEEE DCC*, 1999.

[38] T. Stockhammer and C. Buchner. Progressive texture video streaming for lossy packet networks. *Int. Packet Video Workshop*, 2001.

[39] V. Stankovic, R. Hamzaoui, and Z. Xiong. Packet loss protection of embedded data with fast local search. *IEEE ICIP*, 2002.

[40] R. Albert and A. L. Barabasi. Statistical mechanics of complex networks. *Reviews of Modern Physics*, 74, 47 (2002).

[41] http://www.ilog.com/products/cplex

[42] M. R. Garey and D. S. Johnson. Computers and intractability: A guide to the theory of np-completeness. San Francisco: W. H. Freeman and Company, 1979.

[43] A. V. Goldberg and R. E. Tarjan. A new approach to the maximum-flow problem. *Journal of the ACM*, 35(4):921–940, 1988.

[44] V. A. Vaishampayan. Design of multiple-description scalar quantizers. *IEEE Trans. Inform. Theory*, 39(3):821–834, 1993.

[45] D. Muresan and M. Effros. Quantization as histogram segmentation: globally optimal scalar quantizer design in network systems. In *Proc. Data Compression Conf.*, pp. 302–311, 2002.

[46] H. Jeffreys and B. S. Jeffreys. Dirichlet integrals. In *Methods of Mathematical Physics*, 3d edn. Cambridge: Cambridge University Press, pp. 468–470, 1988.

[47] V. K. Goyal. Multiple description coding: compression meets the network. *IEEE Signal Process. Mag.*, 18(5):74–93, September 2001.

[48] P. Sanders, S. Egner, P. Chou, M. Effros, S. Egner, K. Jain, and L. Tolhuizen. Polynomial time algorithms for multicast network code construction. *IEEE Trans. Inf. Theory*, 51, 2005.

[49] N. Sundaram, P. Ramanathan, and S. Banerjee. Multirate media stream using network coding. *Proceedings of Allerton 05*, 2005.

[50] X. Wu, M. Shao, and N. Sarshar. Rainbow network flow with network coding. *Proceedings of NetCod 2008*, January 2008.

[51] L. Chen, T. Ho, S. H. Low, M. Chiang, and J. C. Doyle. Optimization based rate control for multicast with network coding. *Proceedings of INFOCOM 2007*, May 2007, pp. 1163–1171.

[52] R. W. Yeung. Multilevel diversity coding with distortion. *IEEE Trans. Inf. Theory*, 41(2):412–422, March 1995.

[53] Y. Wu. http://ip.hhi.de/imagecom_g1/savce/downloads/svc-reference-software.htm

[54] H. Schwarz, D. Marpe, and T. Wiegand. Overview of the scalable video coding extension of the h.264/avc standard. *IEEE Transactions on Circuits and Systems for Video Technology, Special Issue on Scalable Video Coding*, 17(9):1103–1120, 2007.

[55] D. Schonberg, K. Ramchandran, and S. S. Pradhan. Distributed code constructions for the entire Slepian-Wolf rate region for arbitrarily correlated sources. *IEEE DCC*, 2004, pp. 292–301.

[56] Y. Matsunaga and H. Yamamoto. A coding theorem for lossy data compression by LDPC codes. *IEEE Trans. Inform. Theory*, 49, 2225–2229, 2003.

[57] R. Yeung, S.-Y. Li, and N. Cai. *Network Coding Theory (Foundations and Trends in Communications and Information Theory)*. Delft, the Netherlands: Now Publishers, 2006.

[58] T. Ho and D. Lun. *Network Coding: An Introduction*. Cambridge: Cambridge University Press, 2008.

[59] B. Rimoldi and R. Urbanke. Asynchronous Slepian-Wolf coding via source-splitting. *IEEE Int. Symp. Inf. Theory*, p. 271, 1997.

[60] T. M. Cover and J. A. Thomas. *Elements of Information Theory*. New York: Wiley, 1991.

[61] A. Grand, B. Rimoldi, R. Urbanke, and P. A. Whiting. Rate-splitting multiple access for discrete memoryless channels. *IEEE Trans. on Inform. Theory*, 47, no. 3, 873–890, 2001.

[62] Y. Cassuto and J. Bruck. Network coding for non-uniform demands. IEEE ISIT, 2005.

[63] T. P. Coleman, A. H. Lee, M. Medard, and M. Effros. A new source-splitting approach to the Slepian-Wolf problem. *IEEE ISIT*, p. 332, 2004.

[64] T. P. Coleman, A. H. Lee, M. Medard, and M. Effros. Low-complexity approaches to Slepian-Wolf near-lossless distributed data compression. *IEEE Trans. Inform. Theory*, 52, 3546–3561, 2006.

[65] S. S. Pradhan, K. Ramchandran, and R. Koetter. A constructive approach to distributed source coding with symmetric rates. *IEEE Int. Symp. Inf. Theory*, 178, 2000.

[66] V. Stankovic, A. Liveris, Z. Xiong, and C. Georghiades. Design of Slepian-Wolf codes by channel code partitioning. *Proc. Data Comp. Conf.*, Snowbird, UT, 2004, pp. 302–311.

[67] Christina Fragouli and Emina Soljanin. Information flow decomposition for network coding. *IEEE Trans. on Inform. Theory*, 52(3): 829–848, 2006.

[68] M. Sartipi and F. Fekri. Distributed source coding in wireless sensor networks using LDPC coding: the entire Slepian-Wolf rate region. *IEEE Wireless Communications and Networking Conference*, pp. 1939–1944, Mar. 2005.

[69] A. D. Wyner. Recent results in the Shannon theory. *IEEE Trans. Inform. Theory*, IT-20, pp. 2–10, Jan. 1974.

[70] A. Aggarwal, M. Klave, S. Moran, P. Shor, and R. Wilber. Geometric applications of a matrix-searching algorithm. *Algorithmica*, 2, pp. 195–208, 1987.

[71] A. Aggarwal, B. Schieber, and T. Tokuyama. Finding a minimum-weight k-link path in graphs with the concave monge property and applications. *Discrete & Computational Geometry*, 12, pp. 263–280, 1994.

[72] A. Albanese, J. Blomer, J. Edmonds, M. Luby, and M. Sudan. Priority encoding transmission. *IEEE Trans. Inform. Theory*, 42, pp. 1737–1744, Nov. 1996.

[73] A. Apostolico and Z. Galil. (eds.). *Pattern Matching Algorithms*, New York: Oxford University Press, 1997.

[74] T. Y. Berger-Wolf and E. M. Reingold. Index assignment for multichannel communication under failure. *IEEE Trans. Inform. Theory*, 48(10):2656–2668, Oct. 2002.

[75] S. N. Diggavi, N. J. A. Sloane, and V. A. Vaishampayan. Asymmetric multiple description lattice vector quantizers. *IEEE Trans. Inform. Theory*, 48(1):174–191, Jan. 2002.

[76] S. Dumitrescu, X. Wu, and Z. Wang. Globally optimal uneven error-protected packetization of scalable code streams. *IEEE Trans. Multimedia*, 6(2):230–239, Apr. 2004.

[77] S. Dumitrescu and X. Wu. Optimal two-description scalar quantizer design. *Algorithmica*, 41(4):300, 269–287, Feb. 2005.

[78] S. Dumitrescu and X. Wu. On global optimality of gradient descent algorithms for fixed-rate scalar multiple description quantizer design. *Proc. IEEE DCC'05*, pp. 388–397, March 2005.

[79] S. Dumitrescu, X. Wu, and Z. Wang. Efficient algorithms for optimal uneven protection of single and multiple scalable code streams against packet erasures. *IEEE Trans. Multimedia*, 9(7):1466–1474, Nov. 2007.

[80] S. Dumitrescu. Speed-up of encoder optimization step in multiple description scalar quantizer design. *Proc. of IEEE Data Compression Conference*, pp. 382–391, March 2008, Snowbird, UT.

[81] S. Dumitrescu and X. Wu. Lagrangian optimization of two-description scalar quantizers. *IEEE Trans. Inform. Theory*, 53(11):3990–4012, Nov. 2008.

[82] S. Dumitrescu and X. Wu. On properties of locally optimal multiple description scalar quantizers with convex cells, to appear in *IEEE Trans. on Inform. Theory*.

[83] M. Effros and L. Schulman. Rapid near-optimal VQ design with a deterministic data net. *Proc. ISIT'04*, pp. 298, July 2004.

[84] M. Effros. Optimal multiple description and multiresolution scalar quantizer design. *Information Theory and Applications Workshop '08*, Univesity of California, San Diego, 27 January–1 February 2008.

[85] M. Fleming, Q. Zhao, and M. Effros. Network vector quantization. *IEEE Trans. Inform. Theory*, 50(8), Aug. 2004.

[86] M. Garey, D. S. Johnson, and H. S. Witsenhausen. The complexity of the generalized lloydcmax problem. *IEEE Trans. Inform. Theory*, 28(2):255–266, Mar. 1982.

[87] V. K. Goyal, J. A. Kelner, and J. Kovačević. Multiple description vector quantization with a coarse lattice. *IEEE Trans. Inform. Theory*, 48, pp. 781–788, Mar. 2002.

[88] S. P. Lloyd. Least squares quantization in PCM. *IEEE Trans. Inform. Theory*, IT-28, pp. 129–137, Mar. 1982.

[89] A. E. Mohr, E. A. Riskin, and R. E. Ladner. Unequal loss protection: graceful degradation over packet erasure channels through forward error correction. *IEEE Journal on Selected Areas in Communication*, 18(7):819–828, Jun. 2000.

[90] D. Muresan and M. Effros. Quantization as histogram segmentation: optimal scalar quantizer design in network systems. *IEEE Trans. Inform. Theory*, 54(1):344–366, Jan. 2008.

[91] J. Østergaard, J. Jensen, and R. Heusdens. n-channel entropy-constrained multiple-description lattice vector quantization. *IEEE Trans. Inform. Theory*, 52(5):1956–1973, May 2006.

[92] R. Puri and K. Ramchandran. Multiple description source coding through forward error correction codes. *Proc. 33rd Asilomar Conference on Signals, Systems, and Computers*, vol. 1, California, Oct. 1999, pp. 342–346.

[93] D. G. Sachs, R. Anand, and K. Ramchandran. Wireless image transmission using multiple-description based concatenated codes. *Proc. SPIE 2000*, vol. 3974, pp. 300–311, Jan. 2000.

[94] S. D. Servetto, V. A. Vaishampayan, and N. J. A. Sloane. Multiple description lattice vector quantization. *IEEE Proc. Data Compression Conf.*, Mar. 1999, pp. 13–22.

[95] V. Stankovic, R. Hamzaoui, and Z. Xiong. Efficient channel code rate selection algorithms for forward error correction of packetized multimedia bitstreams in varying channels. *IEEE Trans. Multimedia*, 14(2):240–248, Apr. 2004.

[96] C. Tian, S. Mohajer, and S. Diggavi. On the symmetric Gaussian multiple description rate-distortion function. *IEEE DCC*, Snowbird, UT, March 2008, pp. 402–411.

[97] C. Tian and S. Hemami. Sequential design of multiple description scalar quantizers. *IEEE Data Compression Conference*, pp. 32–42, 2004.

[98] V. A. Vaishampayan and J. Domaszewicz. Design of entropy-constrained multiple-description scalar quantizers. *IEEE Trans. Inform. Theory*, 40(1):245–250, Jan. 1994.

[99] V. A. Vaishampayan, N. J. A. Sloane, and S. D. Servetto. Multiple description vector quantization with lattice codebooks: Design and analysis. *IEEE Trans. Inform. Theory*, 47(5):1718–1734, July 2001.

[100] J. Barros and S. D. Servetto. Network information flow with correlated sources. *IEEE Trans. Inform. Theory*, 52, pp. 155–170, Jan. 2006.

[101] T. S. Han and K. Kobayshi. A unified achievable region for a general class of multiterminal source coding systems. *IEEE Trans. Inform. Theory*, 26, pp. 277–288, May 1980.

[102] M. Effros, M. Mdard, T. Ho, S. Ray, D. Karger, and R. Koetter. Linear network codes: A unified framework for source, channel and network coding. *Proc. DIMACS Workshop Networking Information Theory*, Piscataway, NJ, 2003.

[103] T. Ho, M. Medard, R. Koetter, D. R. Karger, M. Effros, J. Shi, and B. Leong. A random linear network coding approach to multicast. *IEEE Trans. Inform. Theory*, 53, pp. 4413–4430, Oct. 2006.

[104] A. Ramamoorthy, K. Jain, P. A. Chou, and M. Effros. Separating distributed source coding from network coding. *IEEE Trans. Inform. Theory*, 52, pp. 2785–2795, June 2006.

[105] P. Tan, K. Xie, and J. Li. Slepian-Wolf coding using parity approach and syndrome approach. *Proc. CISS*, pp. 708–713, March 2007.

[106] T. Berger, Z. Zhang, and H. Viswanathan. The CEO problem. *IEEE Trans. Inform. Theory*, 42, pp. 887–902, May 1996.

[107] T. Berger and R. Yeung. Multiterminal source coding with encoder breakdown. *IEEE Trans. Inform. Theory*, 35, pp. 237–244, Mar. 1989.

[108] J. Chen and T. Berger. Robust distributed source coding. *IEEE Trans. Inform. Theory*, 54, pp. 3385–3398, Aug. 2008.

[109] I. Csiszar. The method of types. *IEEE Trans. Inform. Theory*, 44, pp. 2505–2523, Oct. 1998.

[110] I. Csiszar and J. Korner. *Information Theory: Coding Theorems for Discrete Memoryless Systems*. New York: Academic, 1981.

[111] R. L. Dobrushin and B. S. Tsybakov. Information transmission with additional noise. *IRE Trans. Inform. Theory*, 18, pp. 293–304, 1962.

[112] A. Kaspi and T. Berger. Rate-distortion for correlated sources with partially separated encoders. *IEEE Trans. Inf. Theory*, 28, pp. 828–840, Nov. 1982.

[113] Z. Liu, S. Cheng, A. D. Liveris, and Z. Xiong. Slepian-Wolf coded nested lattice quantization for Wyner-Ziv coding: high-rate performance analysis and code design. *IEEE Trans. Inform. Theory*, 52, pp. 4358–4369, Oct. 2006.

[114] Y. Oohama. Gaussian multiterminal source coding. *IEEE Trans. Inform. Theory*, 43, pp. 1912–1923, Nov. 1997.

[115] Y. Oohama. The rate-distortion function for the Quadratic Gaussian CEO problem. *IEEE Trans. Inform. Theory*, 44, pp. 1057–1070, May 1998.

[116] Y. Wang, M. T. Orchard, and A. R. Reibman. Multiple description image coding for noisy channels by pairing transform coefficients. *Proc. IEEE Workshop on Multimedia Signal Processing*, Princeton, NJ, pp. 419–424, June 1997.

[117] V. K. Goyal and J. Kovačević. Optimal multiple description transform coding of Gaussian vectors. *Proc. IEEE Data Compression Conf.*, Snowbird, UT, pp. 388–397, Mar.–April 1998.

[118] V. K. Goyal and J. Kovačević. Generalized multiple descriptions coding with correlating transforms. *IEEE Trans. Inform. Theory*, 47(6):2199–2224, Sep. 2001.

[119] V. K. Goyal, J. Kovačević, and M. Vetterli. Multiple description transform coding: robustness to erasures using tight frame expansions. *Proc. IEEE Int. Symp. Inform. Theory*, pp. 408, Aug. 1998.

[120] V. K. Goyal, J. Kovačević, and M. Vetterli. Quantized frame expansions as source-channel codes for erasure channels. *Proc. IEEE Data Compression*, pp. 326–335, Mar. 1999.

[121] P. A. Chou, S. Mehrota, and A. Wang. Multiple description decoding of overcomplete expansions using projections onto convex sets. *Proc. IEEE DCCn*, pp. 72–81, Mar. 1999.

[122] V. K. Goyal, M. Vetterli, and N. T. Thao. Quantized overcomplete expansions in $\mathbb{R}^N$. *Proc. IEEE Trans. Inform. Theory*, 44, pp. 16–31, Jan. 1998.

[123] S. Rangan and V. Goyal. Recursive consistent estimation with bounded noise. *Proc. IEEE Trans. Inform. Theory*, 47, pp. 457–464, Jan. 2001.

[124] D. Taubman. High performance scalable image compression with EBCOT. *IEEE Trans. Image Processing*, 9, pp. 1158–1170, July 2000.

[125] N. Sarshar and X. Wu. Rate-distortion optimized multimedia communication in networks. *Proceedings of SPIE 08*, volume 6822, January 2008.

[126] X. Wu, B. Ma, and N. Sarshar. Rainbow network flow of multiple description codes. *IEEE Trans. Inf. Theory*, 54(10):4565–4574, Oct. 2008.

[127] S.-Y.R. Li, R.W. Yeung, and N. Cai. Linear Network Coding. *IEEE Trans. Inf. Theory*, 49(2):371–381, Feb. 2003.

[128] N. Gortz and P. Leelapornchai. *Optimization of the index assignments for multiple description vector quantizers. IEEE Trans. Communications*, 51(3):336–340, Mar. 2003.

[129] J. H. Conway and N. J. A. Sloane. *Sphere Packings, Lattices, and Groups*. Berlin: Springer, 1998.

[130] J. Hopcroft and R. Karp. An o($n^{5/2}$) algorithm for maximum matchings in bipartite graphs. *SIAM Journal on Computing*, 2(4):225–231, 1973.

[131] J. Østergaard, J. Jensen, and R. Heusdens. n-channel symmetric multiple-description lattice vector quantization. *Proc. IEEE Data Compression Conf.*, Mar. 2005, pp. 378–387.

[132] X. Huang and X. Wu. Optimal index assignment for multiple description lattice vector quantization. *Proc. of DCC 2006*.

[133] S. Tavildar, P. Viswanath, and A. B. Wagner. The Gaussian many-help-one distributed source coding problem. *IEEE ITW*, pp. 596–600, 2006.

[134] H. Viswanathan and T. Berger. The quadratic Gaussian CEO problem. *IEEE Trans. Inform. Theory*, 43, pp. 1549–1559, Sept. 1997.

[135] A. B. Wanger, S. Tavildar, and P. Viswanath. Rate region of the quadratic Gaussian two-encoder source-coding problem. *IEEE Trans. Inf. Theory*, 54, pp. 1938–1961, May, 2008.

[136] H. S. Witsenhausen. Indirect rate distortion problems. *IEEE Trans. Inform. Theory*, 26, pp. 518–521, Sept. 1980.

[137] A. D. Wyner and J. Ziv. The rate-distortion function for source coding with side information at the decoder. *IEEE Trans. Inform. Theory*, IT-22, pp. 1–10, Jan. 1976.

[138] M. Shao, S. Dumitrescu, and X. Wu. Toward the optimal multirate multicast for lossy packet network. *Proceedings of ACM Multimedia*, 27–31 October 2008.

[139] R. Zamir, S. Shamai, and U. Erez. Nested linear/lattice codes for structured multiterminal binning. *IEEE Trans. Inform. Theory*, 48, pp. 1250–1276, Jun. 2002.

[140] R. Cristescu, B. Beferull-Lozano, and M. Vetterli. Networked Slepian-Wolf: theory, algorithms and scaling laws. *IEEE Trans. Inform. Theory*, 51, pp. 4057–4073, Dec. 2005.

[141] S. Dumitrescu and T. Zheng. Improved multiple description framework based on successively refinable quantization and uneven erasure protection. *Proc. of IEEE Data Compression Conference*, pp. 514–514, March 2008, Snowbird, UT.

[142] S. Dumitrescu, G. Rivers, and S. Shirani. Unequal erasure protection technique for scalable multi-streams. *IEEE Transactions on Image Processing*, 19(2):422–434, Feb. 2010.

[143] S. Dumitrescu, M. Shao, and X. Wu. Layered multicast with inter-layer network coding. *Proceedings of the IEEE INFOCOM*, pp. 442–449, 2009.

Index

3-colorable, 60

H.264, 154
multi-terminal, 169
network coding, 143

Average Normalized Rate, 153

B-MDC, 125–128
Berger-Tung bound, 166
binary symmetric channel, 31, 33
bit allocation, 118, 124

CEO problem, 164, 166–168
channel code, 30, 31
codebook, 91, 93, 161
commodity flow, 11, 20, 48
convex, 29, 41, 45, 47, 53, 54, 58, 77, 86, 88, 117,
      118, 122–124, 127–129, 162
coset, 31, 97, 98, 109, 110
CRNF, 50, 54, 55, 58–63, 67–70

DAG, 50, 54, 55, 58–62
delayed transmission, 143
distortion-rate, 40, 46, 168
distributed source coding, 1, 2, 4, 5, 25, 26, 29,
      35, 170
dummy nodes, 31, 63, 137, 141, 150
dynamic programming, 50, 53, 59, 64,
      118, 121

edge-disjoint paths, 136–138, 149, 150
entropy, 4, 25, 28, 32, 35, 86, 91, 94, 96, 101, 102,
      105, 106, 111, 160
erasure correction codes, 51, 82

fidelity, 16, 61, 68, 73, 117–119, 122–124, 126–129,
      141, 147, 153, 155
Fine Granularity Scalability, 47, 53
finite alphabet, 1, 2, 4, 25
fractional routing, 18, 19, 48, 71, 81

G-MDC, 125–130
Gaussian, 44, 45, 47, 83, 84, 162, 165–167, 170

geometrically similar, 93, 99, 111–113
global coding kernel and vector, 151, 152

Hamming, 31, 162

image, 39, 47, 52, 53, 82, 83
index assignment, 83, 91, 92, 96–102, 105, 108,
      111, 113–116
integer programming, 61, 129
inter-layer, 134, 140, 142, 143, 146, 152, 155
intra-layer, 134, 135, 138, 140, 142, 143, 145–147,
      150, 152–155

joint network source code, 5

Lagrangian, 9, 44, 86, 118, 124
layered, 6, 7, 11, 133–137, 139, 141–143, 146–148,
      151, 153–156
LDPC, 31, 33
Li and Li's conjecture, 10, 23
linear integer program, 147
linear program, 59, 129
local encoding vector, 151

Markov, 168
matrix search algorithm, 121–123
max-flow, 4–6, 15, 16, 35, 39–41, 68, 69, 133–139,
      141, 142, 145, 146, 153, 154, 157
max-SNP Hard, 19
maximum flow, 135
MDC, 1, 7, 10, 39–48, 50–55, 59, 60, 62, 67, 68,
      71–74, 82, 83, 91, 107, 116, 117, 125, 126
MDLVQ, 91–94, 96, 100–102, 105, 107, 108, 110,
      115, 116
MDSQ, 83–90
MDVQ, 84, 91
min-cut, 35, 133
monotonically non-decreasing, 117
monotonically non-increasing, 117–119, 122, 123
multi-terminal, 164–166, 169, 170
multicast layer, 134
multicast session, 134–136
multiple-description, 1, 6, 7, 11, 39, 40, 42–44, 50,
      70, 73, 74, 82–84, 91, 125, 132

NASCC, 7–11, 15, 16, 40–43, 50, 54, 71
nested codes, 163
network coding, 3–11, 15, 16, 19–25, 34–36, 39, 40,
    44, 45, 55, 57, 125, 126, 132–135, 137, 138,
    140–144, 146, 152, 153, 155
network information flow, 3, 8, 15, 39, 133, 157
NP-Hard, 50, 58, 59, 61

packetization, 117–124
parity check matrix, 31, 33, 34, 163
Peer-to-Peer, 7
PET, 46, 47, 50–53, 55, 59, 71–74, 78, 82, 83, 116,
    117, 125–128
polynomial time, 19, 60, 139, 148
progressive image encoders, 52, 53
progressive source code, 47, 51, 132
progressive source coding, 1
PSNR, 127, 141, 147, 154–156

quadratic Gaussian, 162–169

Rainbow Network Flow, 11, 16, 41, 48, 59, 70, 71,
    81, 133
rate-distortion, 1, 2, 4, 9, 45, 47, 52, 53, 78, 83, 86,
    125, 127, 163, 167–169
redundancy assignment, 117–124, 126, 127
Reed Solomon Codes, 51, 83
RNC, 133
RNF, 11, 41, 48–50, 54, 58, 59, 63, 66, 67, 70–75,
    80, 81, 133

routing, 3, 4, 7, 10, 11, 15–18, 20–23, 35,
    39, 41–43, 45, 47, 48, 71, 125, 126,
    132, 133

scalar quantizer, 83, 84, 86
secondary decoders, 170
sensor networks, 7, 25, 36
side information, 11, 158, 159, 162, 166, 168
sink node, 1, 3–6, 8, 9, 15, 17, 18, 22, 34, 35,
    39–43, 45, 48–50, 52, 58, 60, 64, 71, 73, 77,
    78, 125, 132–137, 141, 144, 150, 153,
    155, 158
sink nodes, 132
Slepian-Wolf, 2, 4, 5, 11, 25–27, 29–36, 163,
    164, 169
Steiner tree packing, 18, 19, 41
sublattice, 91–94, 96–102, 107–113, 115
syndrome, 31–34

time-sharing, 18, 29, 30, 32
turbo codes, 31

UEP, 117–124, 142

video, 7, 8, 17, 39, 47, 82, 83, 154

Wyner, 2
Wyner-Ziv, 2, 10, 11, 31, 32, 158–164

XOR, 5, 21, 22, 132, 133

Printed in the United Kingdom by TJ Clays Ltd.